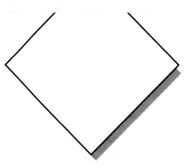

HUMAN ANATOMY AND PHYSIOLOGY LABORATORY MANUAL

WITH PHOTO ATLAS

◆◆◆

SECOND EDITION

STEPHANIE IRWIN
FRONT RANGE COMMUNITY COLLEGE

KENDALL/HUNT PUBLISHING COMPANY
4050 Westmark Drive   Dubuque, Iowa 52002

This project is dedicated to my dad, Tom Clark, whose courage and grace are examples for my life. Also, special thanks to Tom for his photography in Figure 8.2 and Photos 351a, 351b, 356a, 364a, 364b, 367, 374a, 376b, 382, 383, 386a, 386b, 399–408b.

Illustrations by Jamey Garbett. © 2003 Mark Nielsen: *Figures:* I.1, I.2, I.3, I.4, 6.1, 7.1, 7.2, 7.3, 7.4, 7.5, 7.6, 8.1, 8.3, 8.4, 8.5, 8.6, 8.7, 8.8, 8.9, 8.10, 8.11, 8.12, 9.1, 9.2, 9.3, 9.4, 9.5, 9.6, 9.7, 9.8, 9.9, 9.10, 9.11, 9.12, 9.13, 10.1, 10.2, 10.3, 10.5, 10.6, 11.1 (partial), 11.2, 12.1, 12.2, 12.3, 12.4, 12.5, 12.6, 12.7, 12.8, 12.9, 12.10, 12.11, 12.12, 12.13, 12.14, 12.15, 12.16, 12.17, 12.18, 12.19, 12.20, 12.21, 12.22, 12.23, 12.24, 12.25, 12.26, 12.27, 12.28, 12.29, 12.30, 12.31, 12.32, 12.33, 12.34, 12.35, 12.36, 12.37, 12.38, 12.39, 12.40, 12.41, 13.2, 14.2, 14.3, 14.6, 15.1, 15.2, 15.3, 16.1, 16.4, 16.7, 16.8, 20.1, 21.1, 21.3, 21.4, 21.5, 21.6, 21.7, 21.8, 21.9, 21.10, 23.1, 23.2, 23.3, 23.4, 23.5, 25.2, 25.3, 25.5, and 27.3. *Images on pages:* 68, 78, 79, 91, 92, 109, 120, 121, 130, 205, and 305.

CONTENTS

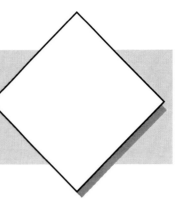

PREFACE

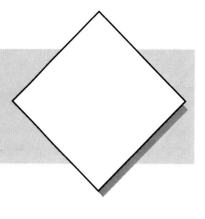

Welcome to the study of Human Anatomy and Physiology! This is an exciting and involved course in which you will examine the gross and microscopic anatomy, histology and physiology of each of the body's organ systems. This laboratory manual is designed to introduce you to critical concepts underlying many allied health programs such as nursing, paramedic, respiratory therapy and dental hygiene. Although a review of cell biology is included, it is important that you have the appropriate preparation for this course to provide you with the best opportunity for success. Completion of college level biology is strongly recommended before taking this class. For appropriately prepared students, the laboratory experience can be an exhilarating journey through the human body.

A Note about Dissection:

Examination of diagrams and anatomical models is an excellent method to begin your study of anatomy and physiology. However, no study of anatomy can be complete without the kinesthetic experience of mammalian dissection. Manipulation of tissues and organs in specimens provides the opportunity for experiencing the relationships of various structures to one another, and for palpating specific landmarks and textures. This process engages different learning operations in your brain than basic observation and memorization, and acts to further increase your understanding of the body as a whole. While dissection and exploration of a human cadaver is the ideal method of discovery, logistical complications prevent most educational institutions from offering such an opportunity. This course utilizes prepared mammalian specimens obtained from biological supply companies to best approximate the human form. As students and instructors of anatomy and physiology, it is important to value life. It is, therefore, with utmost respect and appreciation that we should study the animal specimens that make our learning possible.

LABORATORY GUIDELINES
AND SAFETY

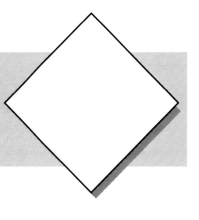

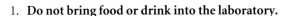

1. **Do not bring food or drink into the laboratory.**

2. If you are pregnant or have a diagnosed medical problem notify the instructor and consult with your physician regarding the possibility of risk to your health in the laboratory environment.

3. Know the location of the safety features in the laboratory (eyewash, fire extinguisher, first aid kit).

4. Anatomy involves dissection of preserved animals or animal parts. You are expected to participate in the dissections and to take any quizzes or practical examinations related to the dissections. Discuss any objections or health problems related to animal dissection with your instructor on the first day of class.

5. You may choose to wear old clothing or bring a lab coat to lab.

6. Place your belongings where they will not get spills on them and where individuals moving in the aisles will not trip over them.

7. Wear disposable gloves for dissections. Remove your gloves when leaving the room for any reason or when handling the microscopes. Use safety glasses at your own discretion. You may also wish to bring a surgical mask if the specimen fixative irritates your nose or throat.

8. Dispose of unnecessary animal specimens or parts in a designated autoclave bag. Gloves and paper towels may be discarded in the regular classroom trash container.

9. If you spill a chemical on your skin immediately flush the area with cool water and notify the instructor. If you get a chemical in your eyes use the eyewash and notify the instructor immediately.

10. If you break glassware or cut yourself during a lab notify the instructor immediately. Handle all dissection instruments with caution.

11. Thoroughly disinfect all work surfaces at the end of all laboratory exercises.

12. Wash your hands at the end of each lab.

INTRODUCTION: THE HUMAN BODY AND ANATOMICAL TERMINOLOGY

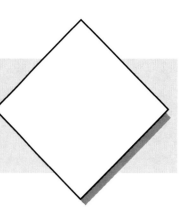

Objectives

1. Define **anatomical position** and use regional terms to locate specific body areas
2. Use directional terms to describe the position of a structure or injury
3. Explain three body planes used to study human anatomy
4. Locate the major body cavities and subcavities, and indicate important organs in each

INTRODUCTION

Communication can be difficult in Anatomy and Physiology if students don't learn the specific anatomical language. It is important that members of anatomical and allied health fields be able to communicate with one another using a logical, standardized language that is descriptive and informative. Otherwise, we may resort to using terms that are confusing or conditional. For example, the nose is "above" the chin only when the head is in an upright position. But if an individual is laying on her side how would you describe the position of the nose? This exercise offers an opportunity to become familiar with common anatomical terminology. You will continue to use this information throughout the remainder of this course, and beyond if you enter training in an allied health career.

CONCEPTS

I. Anatomical Position

We will start with a reference point that will help avoid confusion when describing the positions of body parts, landmarks or injuries. This reference point is a body position that is termed the **anatomical position.** In the anatomical position the human subject is standing upright with toes and eyes facing forward (see Figure I.1). The arms hang to the side with the palms facing forward. You should always describe the position of a structure or injury *as though the subject is in the anatomical position,* even if that individual is turning a cartwheel or standing on her head. Thus, in the example above, the nose is always above the chin, even when a person is upside down!

II. Regional Terminology

Regional terms are descriptive anatomical words used to describe a specific location in the body. We will concentrate on the surface regions shown in Figure I.1. **Axial** refers to the region that includes the head, neck and torso. **Appendicular** refers to the appendages, or the arms and legs.

III. Directional Terminology

Directional terms are descriptive anatomical words that describe the location of a specific structure, body part or injury with respect to a different structure or body part. We can use several different directional terms to describe the same landmark, depending on the body part to which it is being compared. For example, Colorado can be described as being both *north* of New Mexico and *south* of Wyoming. Both *north* and *south* are accurate directional descriptions given the appropriate comparison.

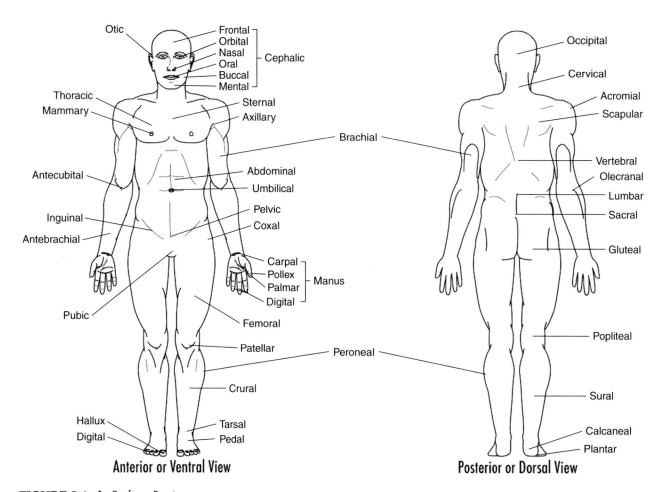

FIGURE I.1 ◆ Surface Regions

The following are common anatomical directional terms:

- **superior**—closer to the top of the head; above (usually used in axial region)
- **inferior**—closer to the bottom of the feet; below (usually used in axial region)
- **anterior** (also called **ventral** in humans)—closer to the front
- **posterior** (also called **dorsal** in humans)—closer to the back
- **medial**—closer to the midline of the body
- **lateral**—farther from the midline of the body
- **intermediate**—in between
- **proximal**—closer to the attachment of the body part to the trunk (used in appendages)
- **distal**—farther from the attachment of the body part to the trunk (used in appendages)
- **superficial**—closer to the surface of the skin; more external
- **deep**—farther from the body surface; more internal

Be aware that the terms *right* and *left* refer to the subject's right and left, not yours.

IV. **Sectional Terminology**

In anatomy we often view a "slice," or plane, through a body. Keep in mind that, while planes through a whole body are easy to identify, those through unfamiliar organs may be more difficult. Read figure labels to understand from which angle you are viewing an organ slice.

- **Midsagittal section**—divides the specimen into equal left and right halves
- **Sagittal section**—divides the specimen into left and right sides, not necessarily equal
- **Frontal section**—divides the specimen into anterior and posterior parts
- **Transverse (cross) section**—divides the specimen into superior and inferior parts

V. **Body Cavities**

The position of organs and structures can often be described using their locations in cavities, or spaces, within the body. The largest body cavities have no opening to the outside of the body. These large cavities can be classified by their position in the dorsal or ventral region of the body.

Dorsal Cavities

- **Cranial Cavity**—located in the skull; contains the brain
- **Vertebral Cavity**—created within the stacked vertebra of the spinal column; encloses the spinal cord

Ventral Cavities

- **Thoracic Cavity**—the chest cavity is bounded posteriorly by the spinal column, laterally and anteriorly by the rib cage, and inferiorly by the diaphragm muscle; contains the lungs and heart
  - Pleural cavities—spaces within the membranes that surround the lungs
  - Mediastinum—the space in between the lungs; contains the heart
  - Pericardial cavity—the space within the membrane that surrounds the heart
- **Abdominopelvic Cavity**—the space bounded superiorly by the diaphragm muscle and inferiorly by the pelvic floor; this cavity is artificially divided into two regions: the abdominal and pelvic cavities
  - Abdominal cavity—superior portion of the abdominopelvic cavity; contains the digestive viscera (organs)
  - Pelvic cavity—the inferior portion of the abdominopelvic cavity located between the pelvic bones; contains the urinary bladder, the rectum and reproductive organs
  - Peritoneal cavity—the space within the membrane that lines the abdominal cavity and digestive viscera

Other body cavities are smaller, and many are open to the outside of the body. These include the oral cavity (mouth), nasal cavity and orbital cavities (orbits), which contain the eyeballs.

Midsagittal

Sagittal

Frontal

Transverse
(cross section)

FIGURE I.2 ◆ Planes

VI. **Membranes**

As we progress through our study of the human body we will encounter different types of membranes that line cavities or organs. For now we can make some general observations about three main types of membranes.

- **The Integument**—the skin; this membrane protects all of the underlying structures and has some metabolic functions
- **Mucous membranes**—line cavities and passageways that are open to the outside of the body; these membranes secrete mucus
- **Serous membranes**—line closed ventral body cavities; these membranes are double layered and have specific names depending on their location:
  - **Pericardium**—surrounds the heart
  - **Pleurae**—surround the lungs; singular is *Pleura*
  - **Peritoneum**—lines the abdominal cavity and the digestive viscera

Serous membranes can be visualized by imagining your fist pressed into a water balloon (Figure I.3). Your fist represents the organ (heart, lung or digestive organ) and the balloon represents the serous membrane. The water within the balloon represents fluid, called **serous fluid,** that is secreted into the space between the membrane layers. As your fist presses into the balloon the balloon "skin" forms two layers; one layer is in contact with your fist and the other is an outer layer. The water fills the space between the two layers.

The inner layer of the serous membrane, (that part of the balloon that touches the fist), lines the organ and is called the **visceral** layer. Thus, the visceral pericardium lines the heart, the visceral pleurae line the lungs, and the visceral peritoneum lines the digestive viscera.

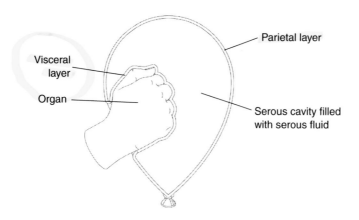

FIGURE I.3 ◆ Water Balloon Analogy for Serous Membranes

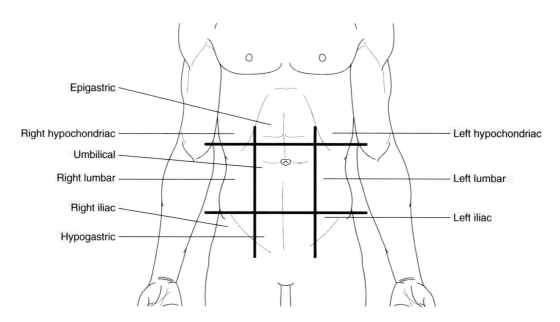

FIGURE I.4 ◆ Abdominopelvic Regions

The outer layer of the serous membrane, (that part of the balloon that does not touch the fist), lines the ventral body cavity and is called the **parietal** layer. Thus, the parietal pericardium lines the mediastinum, the parietal pleurae line the thoracic cavity, and the parietal peritoneum lines the abdominal cavity.

The serous fluid is named for the name of the specific serous membrane that secretes it. The fluid fills the slit-like cavity between the visceral and parietal layers. For example, pericardial fluid fills the pericardial cavity between the visceral and parietal pericardium.

VII. Abdominopelvic Regions

For purposes of further specifying locations of structures or injuries, the abdominopelvic cavity can be subdivided into nine regions. These regions are positioned in three rows of three regions as shown in Figure I.4.

Name _____

Introduction: The Human Body and Anatomical Terminology

◆ Practice

1. Define and describe the **Anatomical Position:**

2. Write the correct anatomical term for each of the regions listed below:

Big toe _____

Elbow _____

Eye _____

Forehead _____

Cheek _____

Armpit _____

Hand _____

Buttock _____

Chin _____

Hips _____

Breasts _____

Wrist _____

Back of knee _____

Back of head _____

Foot _____

Thumb _____

3. Write the appropriate directional term on the lines below to correctly describe the relationships between body parts. Remember to use the anatomical position as your reference point!

The thumb is _____ to the ring finger.

The elbow is _____ to the shoulder.

The skin is _____ to the bone.

The knee_____ to the ankle.

The belly button is _____ to the chin.

The scapular region is _____ to the thoracic region.

The cephalic region is _____ to the sternal region.

The heart is _____ to the lungs.

Bone is _____ to muscle.

The olecranal region is _____ to the antecubital region.

4. Assign each of the following organs to one or more of the body cavities below:

a. brain

b. spinal cord

c. liver

d. spleen

e. large intestine

f. stomach

g. small intestine

h. heart

i. pancreas

j. lungs

k. ovaries/uterus

l. urinary bladder

Dorsal body cavities:

• cranial cavity: contains _____

• spinal cavity: contains _____

Ventral body cavities:

• thoracic cavity: contains _____

• abdominal cavity: contains _____

• pelvic cavity: contains _____

5. Choose which body cavity needs to be opened for each of the surgical procedures listed below (more than one choice may apply):

a. abdominopelvic

b. thoracic

c. pleural cavity

d. cranial

e. pericardial

_____ removal of a diseased lobe of the lung

_____ removal of a brain tumor

_____ operation on the liver

_____ triple bypass surgery on the heart

_____ removal of a segment of the large intestine

6. Choose the appropriate membrane type to match each description below:

    a. integument
    b. serous
    c. mucous

    _____ lines open cavities or passageways

    _____ lines closed ventral cavities

    _____ lines the inside of the stomach

    _____ covers the outer surface of the body

    _____ peritoneum

    _____ contains a fluid-filled cavity between two membrane layers

7. Without looking at a picture, diagram and label the 9 regions of the abdominopelvic cavity.

Right Hypochondriac	Epigastric	Left hypochondriac
Right Lumbar	Umbilical	Left Lumbar
Right Iliac	Hypogastric	Left Iliac

CARE AND USE OF THE MICROSCOPE

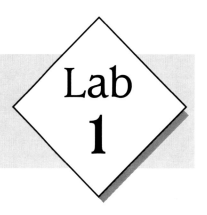

Lab 1

Objectives

1. Identify the parts of the compound microscope and understand their functions
2. Explain how to care for the microscope
3. Demonstrate the ability to correctly use the microscope in preparation for future lab exercises

Materials

- Compound microscopes
- Immersion oil
- Lens paper
- Lens cleaning solution
- Microscope slides
  - Letter 'e' slides
  - Various tissue samples

INTRODUCTION

Anatomy and Physiology involves the study of both gross anatomy and microscopic anatomy. Gross anatomy includes those structures we can see with the unaided eye. Microscopic anatomy requires magnification for our eyes to be able to view. Throughout this course we will be studying microscopic anatomy in addition to the gross anatomy of each organ system. It is therefore important that you have a basic familiarity of how to care for and use the microscopes in the lab.

The microscopes we use in this class are compound microscopes. A compound microscope magnifies the image of the specimen you are viewing twice. It may also be referred to as a binocular microscope because it has two eye pieces, or ocular lenses. The image you see is of the specimen on a brightly lit background, so these microscopes are also classified as brightfield microscopes.

CONCEPTS

I. Parts of the Microscope

The pathway of light begins at the light source in the base of your microscope. Light passes through a **condenser,** which focuses the light waves through an opening called the **iris diaphragm** and then through the microscope slide. The condenser adjustment knob moves the condenser up and down to improve detail while the iris lever attached to the condenser opens and closes the iris to allow more or less light to pass. More light allows greater visibility but less detail and contrast in the image of the specimen.

The microscope slide sits atop the **stage** and is held in place by a spring-loaded moveable arm. The **mechanical stage** mechanism has two knobs that enable the mechanism to move both forward and backward, and side to side.

Light passes through the specimen and into the first lens that magnifies the image, the **objective** lens. Your microscope has four objective lenses, each with a different power of magnification: 4X, 10X, 40X, 100X. A **revolving nosepiece** allows you to move any of the four objectives into place above the specimen. From the objective lens, light continues into the body of the microscope and is reflected off of a mirror to the second set of lenses, the **ocular** lenses. These are located in the eyepieces through which you observe the specimen. The ocular lenses have a 10X power magnification in your microscopes.

The image of the specimen is controlled by manipulating specific parts of the microscope. At one side of the base is located a slider switch that controls the brightness of your **light source.** You will want to adjust the amount of light to best view the image of the specimen. The image is focused by both the **coarse focus knob** and the **fine focus knob.** The coarse focus knob moves the stage up and down to help you bring the image into a rough focus. The fine focus knob also moves the stage up and down, but requires many more turns for a small degree of movement. This allows for precision focusing. Again, the iris lever controls the diameter of the iris diaphragm within the condenser, and therefore the amount of light that passes through the microscope slide. Finally, one of the oculars also focuses the image for one of your eyes.

◆ Exercise 1.1 *Review Parts of the Microscope*

Use Photo 351a and Part I to identify the following structures and describe their functions.

Ocular:

Objective:

Revolving Nose Piece:

Stage:

Mechanical Stage:

Coarse Focus Knob:

Fine Focus Knob:

Iris Diaphragm and Lever:

Condenser:

Light Source:

Base:

Arm:

––––––––––––––––––––––––––––––––––––– ◈◈◈ –––––––––––––––––––––––––––––––––––––

II. Care of the Microscope

Quality microscopes are sophisticated and expensive instruments. In order to avoid costly maintenance and repairs it is extremely important that you diligently care for your microscope using the following instructions:

1. Always use the same microscope for your lab partner and yourself.
2. Use care not to bump the microscope on the sides of the storage cabinet when removing it. **Carry your microscope in two hands.** One hand should support the base and the other should grasp the arm of the microscope. Do not be tempted to carry a microscope in each hand. Also, be careful to make sure the cord is wrapped up when you are carrying the scope. Trailing cords can trip you or others and lead to injuries and broken scopes.
3. When you reach the table, **gently** set the microscope down with the arm facing away from you. The oculars should face you. **Never slide the microscope across the table.** Vibrations caused by sliding not only change the focus, but loosen screws and lenses. If you need to change the microscope position, carefully pick it up to move it.
4. Never use tissue or your shirt to clean the objectives and oculars. Always use specially designated lens paper.
5. Follow this sequence when preparing to put away your microscope:
   • Clean all lenses and the stage with lens paper. Clean the oil immersion lens (100X objective) last. Make sure all oil has been removed.
   • Click the lowest power objective (4X) into place above the stage. Make sure no microscope slide is on the stage.
   • Turn the coarse focus knob so that the stage is as far away from the objective as possible.
   • The oculars should remain facing toward the stage. If they are loose and able to move backward and forward, reposition them and tighten the screw at the base of the oculars on the right side.
   • Coil the cord and secure it with a rubber band. Do not wrap the cord around the scope.
   • Cover the scope and return it to the storage cabinet, being careful not to bump it as you set it in its numbered space.

RECORD THE NUMBER OF YOUR MICROSCOPE HERE _____

III. Magnification and Resolving Power

The total magnification of the image you observe through the oculars is a product of the objective lens magnification and the ocular magnification. Thus you would use this formula to determine the total magnification:

Total Magnification = ocular magnification × objective magnification

The power of each objective lens is printed on the side of the objective. See Photo 351b.

◆ **Exercise 1.2** *Determine Total Magnification*

Calculate the total magnification for each of the objective lenses.

Objective Power	Ocular Power	Total Magnification
1.		
2.		
3.		
4.		

IV. Using the Microscope

◆ **Exercise 1.3** *Set Up the Microscope and View a Specimen*

A. Obtain the following materials:
1. Bottle/tube of immersion oil
2. Lens paper
3. Microscope slide with the letter 'e'
4. One or two microscope slides with various tissue types

B. Set up the microscope and view a tissue specimen by following these steps:
1. Turn on the light on the side of the base and slide light switch to the dimmest position, toward the back of the scope. Later you may increase light levels as needed.
2. Move the condenser to the highest position.
3. Using the revolving nosepiece, click the 4X objective in place above the stage. *Always view a new slide starting at the lowest power objective.*
4. Look through the oculars. Notice that the oculars slide toward each other. Adjust the distance between the oculars so that you are seeing only one bright circle, or field. It is important to *keep both eyes open*. Notice that one of the oculars has a pointer in it. You can determine which ocular the pointer is in by closing one eye at a time.

5. Using the coarse focus knob, make sure that the stage is as far from the objective as possible. Place a microscope slide with a cell or tissue type onto the stage.

   ▶ Which objective should be in place over the stage when you first view every microscope slide? _____

6. Look through the oculars. Increase the amount of light if necessary. Using the coarse focus knob move the stage closer to the objective lens until you achieve the best focus. Then, use the fine focus to clarify the image. Note: Your microscope has a property called *parfocal*. This means that each time you move a higher power objective into place the image remains roughly in focus. Therefore *you should only use the FINE focus as you increase magnification*. Using the coarse focus at 40X objectives or higher can cause the objective lens to scrape the slide, scratching the lens and potentially breaking the slide.

   (Having trouble focusing? Make sure the microscope slide is label-side up.)

7. When you have achieved the best focus using the fine focus adjustment, close your eye on the same side with the ocular that has a focus adjustment. Adjust the fine focus adjustment again, if necessary. Now close the opposite eye and use the ocular focus ring to improve clarity for that eye. This procedure allows for differences in vision between your two eyes.

8. While still using the 4X objective, move the iris lever back and forth and the condenser knob up and down.

   ▶ What is the effect of the iris diaphragm?

   ▶ What is the effect of the condenser movement?

   (If these are unclear, try them again when you move to a higher magnification.)

   Pick out some landmark structures around the edge of the field of view before proceeding.

9. After both lab partners have had a chance to practice setting up the microscope, move the next highest power objective lens (10X) into place above the slide.
   Compare the following with the 4X objective. You may have to move back and forth between lenses.
      The **working distance** is the distance between the objective lens and the top surface of the microscope slide.

   ▶ Is this distance more or less? _____

   Do you see more or less of the specimen in the field of view? _____
   Hint: Can you still see the landmark structures you picked out in step 8?

   ▶ Do you see more or less detail in the specimen? _____

   Repeat these observations using the 40X objective.

   ▶ Should you use the coarse focus adjustment to bring the image back into focus? _____

   Record your observations of working distance and specimen detail here:

Practice repeating the steps of this exercise with different microscope slides.

Always focus each objective lens in order. Do not skip lenses or you will find it difficult to focus at higher powers. Also, only use the coarse focus at 4X and 10X magnifications.

C. **Understanding the orientation of the image**

Controlling the microscope slide may be confusing until you gain familiarity with how your manipulations affect the image you see.

1. Move the 4X objective into place and remove the microscope slide. Observe the letter 'e' slide without the use of the microscope.

   Draw the 'e' as you see it with your unaided eye:

2. Using the procedure you have just learned, place the 'e' microscope slide on the stage and view at 10X. (Remember to focus at 4X first.)

   Draw the 'e' as you see it in the microscopic field of view:

   ▶ How does the image of the specimen in the microscopic field of view compare to the actual specimen on the microscope slide?

3. Using the mechanical stage, move the slide to the right. Which direction does the image move?

4. Using the mechanical stage, move the slide away from you. Which direction does the image move?

D. **Use of the Oil Immersion (100X) Lens**

When you want to increase the magnification and detail of the image of an object, you will use the oil immersion lens. Since this objective has the smallest diameter lens, it provides the greatest resolution but lets in less light. In order to prevent light rays from scattering, and to direct more light into the small 100X lens, oil is used to decrease light refraction. Please note that excess oil on the slide can enter the objective and require expensive cleaning or replacement. So *use oil sparingly!*

1. Move the 4X objective into place above the stage. Choose a slide with a tissue specimen and successively focus the image at each objective up to 40X. Without changing the focus, move the 4X objective back into place over the slide.

2. Using a bottle of immersion oil, place *one* drop of oil directly on top of the specimen. Now, *without changing the focus,* swing the 100X lens into place above the slide. The small glass lens on the objective will enter the drop of oil but should not touch the slide if you have not moved the coarse focus.

3. Using the **fine focus adjustment,** bring the image into focus. You may have to increase the light or open the iris diaphragm.

4. Once you have the image in focus, move the iris lever back and forth and the condenser knob up and down.

▶ What happens to the detail and contrast of the image when you decrease the light?

▶ How does moving the condenser affect the image?

▶ Why is it important not to move the course focus when using the 100X objective?

5. After both lab partners have had a chance to set up an oil immersion image you need to clean up the oil. Always clean the oil immersion lens and the microscope slide with lens paper. *DO NOT* use the oily lens paper on the other objectives. Always use a clean piece of paper for non-oil lenses.

After you have completed these exercises, and feel comfortable with the use of the microscope, follow step 5 under "Care of the Microscope" to put the scope away.

◆◆◆

ORGAN SYSTEMS OVERVIEW—FETAL PIG DISSECTION

Lab 2

Objectives

1. Examine internal pig anatomy
2. Classify structures in organ systems
3. Learn general functions of observed structures
4. Gain an appreciation of the complexity and elegance of function of the body systems

Materials

- Fetal pigs
- Dissecting tray
- Dissecting instruments
- String

INTRODUCTION

The organ systems of the body all work together to operate a functioning organism. Although we study the organ systems one at a time in this course, you need to understand that the systems are intricately interconnected. The traditional method of study for anatomy and physiology is to address each system separately and to learn the primary functions of each. However, if you first consider a specific process, you may be better able to discern the interrelated activities of all of the systems together. Take cellular metabolism, for example. Body cells must have oxygen and nutrients to perform cellular metabolism. The oxygen is delivered to the blood through the *respiratory system,* and nutrients are absorbed into the blood by the *digestive system.* Both oxygen and nutrients are transported via the *cardiovascular system* to the various body tissues. During metabolism cells produce waste products, including carbon dioxide. This waste gas is eliminated through the *cardiovascular* and *respiratory systems,* while other waste products are eliminated through the *urinary system.*

This is just a quick snapshot of the cooperation between several organ systems to accomplish one function. In this lab you will observe the organs of body organ systems in a fetal pig.

CONCEPTS

This lab is intended to provide an overview of organ systems as an introduction to human anatomy and physiology. Although directional terminology is different when describing quadruped animals, we are using the pig as a model for *human* anatomy and so will apply human directional terms to the pig to avoid confusion. Note that the pig's arteries and veins have been injected with red and blue latex for ease of identification.

◆ **Exercise 2.1** *Dissect and Examine a Fetal Pig*

Refer to Photo 352 as you proceed through this exercise.

A. **Place your pig on its back and tie its legs apart by tying a string to one leg and wrapping it around the dissecting pan before tying it to the other leg.**

B. **Head and Neck Region**

1. Paired **external nares** at the anterior snout are openings into the respiratory system. Air from the nares passes posteriorly and inferiorly into the **pharynx.** Inferior to the pharynx is the **larynx** that contains a small opening into the **trachea.** A small flap of cartilage, the **epiglottis,** covers the opening when the pig swallows to prevent food from entering the trachea.

2. The **thyroid gland,** part of the endocrine system, lies anterior to the laryngeal region. It functions in regulating metabolism. Another endocrine gland located anterior to the trachea is the **thymus,** which functions in the immune system.

C. **Thoracic Region**

1. Beginning superior to the umbilical cord, make a shallow midline incision and continue it superiorly to the inferior aspect of the sternum. Use scissors to continue the incision superiorly up the sternum. From the midline incision make two incisions extending laterally at both the superior and inferior boundaries of the thoracic cavity. Open the rib cage to identify the thoracic structures.

2. Note the centrally located **heart,** a major component of the cardiovascular system. Projecting from the superior aspect of the heart are the **great vessels,** which include the aorta and the superior vena cava. The heart may still be covered by a serous membrane called the **pericardium.**

3. Lateral to the heart are the **lungs,** a continuation of the respiratory system. The lungs are connected to the trachea via two branches called the **primary bronchi,** which exit the lungs posterior to the heart.

4. Posterior to the trachea lies the **esophagus,** a transport tube of the digestive system that delivers food to the stomach.

D. **Abdominal Region**

1. Continue the midline abdominal incision inferiorly to one side of the umbilical cord. Make two lateral incisions to allow access to the abdominopelvic cavity.

2. Note the flat muscle that separates the thoracic cavity from the abdominal cavity. This **diaphragm** muscle is responsible for normal inspiration of air into the lungs.

3. Inferior to the diaphragm is the large, multilobed **liver.** The liver is an accessory organ of the digestive system that acts to detoxify blood after it leaves the intestines and before it is returned to the heart. It also makes bile that is stored in the **gallbladder** nestled between the lobes of the liver.

4. Find the sac-like **stomach** inferior to the liver. The esophagus joins the stomach after passing through the diaphragm. Find the dark brown **spleen,** another organ that filters blood. The spleen is located inferior to the stomach, and extends from the left side.

5. Food passes from the stomach into the **small intestine.** Attached to the inferior aspect of the stomach is a large, specialized serous membrane called the **omentum.** The omentum will have to be lifted to view the small and large intestines. Other specialized serous membranes, the **mesenteries,** secure the small intestines in place within the abdominal cavity, and provide routes for blood vessels and nerve supplies.

6. The small intestine is continuous with the **large intestine.** The large intestine allows undigested material to pass into the **rectum** and out of the body.

7. Another accessory organ of the digestive system is the subtle tan **pancreas.** The pancreas is located inferior and slightly lateral to the stomach, deep to the omentum. It secretes digestive juices into the small intestine. It also acts as an endocrine organ that helps to regulate blood sugar levels.

8. The two large bean-shaped **kidneys** on either side of the spinal column are part of the urinary system. They secrete urine into thin, muscular **ureters,** which transport the urine to the urinary bladder. From the **urinary bladder** urine is excreted through the urethra.

9. For the reproductive system in a female pig, locate the small, pale **ovaries** inferior to the kidneys. Coiled **uterine tubes** transport eggs from the ovaries to the paired horns of the V-shaped uterus. Human females have a single chambered uterus. The inferior region of the uterus is the cervix, which leads to the **vagina.** The urethra lies anterior to the vagina.

If your pig is a male, locate the two undescended **testes** anterior to the anus. The **penis** contains the urethra.

Note: Be sure to examine both a male and a female pig.

E. **Cleanup**

Place all tissue into a trash bag reserved for this purpose. Gloves and paper towels should be discarded in regular trash cans. Rinse dissecting instruments and return them to their storage containers. Rinse dissecting trays and dry them before returning them to their storage location. Spray your table surface with disinfectant and wipe it clean.

Name _____

Organ Systems Overview

◆ **Practice**

1. Match the appropriate organ system with each of the physiological functions listed below. You may need to refer to your text.

   a. integumentary system
   b. skeletal system
   c. muscular system
   d. nervous system
   e. endocrine system
   f. cardiovascular system
   g. lymphatic system
   h. respiratory system
   i. digestive system
   j. urinary system
   k. reproductive system

   _____ secrete hormones to regulate homeostatic processes

   _____ return fluid from the tissues back to the blood circulation

   _____ respond to stimuli and stimulate a fast response by muscles or glands

   _____ regulate fluid balance, electrolyte balance and acid-base balance

   _____ allow movement of the skeleton and of materials within hollow organs

   _____ covers and protects deeper tissues from physical and chemical trauma, and from infection

   _____ supplies blood with oxygen and removes carbon dioxide from the blood

   _____ transports blood and the contained nutrients, proteins and respiratory gases

   _____ produce cells that combine to form offspring; protect and nurture offspring

   _____ physically and chemically break down food into components that are absorbed into the blood

   _____ allow movement of the organisms when acted upon by muscles; protects vital organs

2. For each of the pairs of organ systems below, list functions in which both are involved and cooperate together.

   Skeletal and Muscular systems

Cardiovascular and Digestive Systems

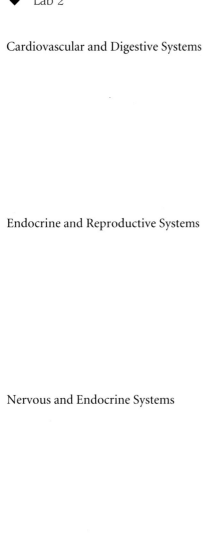

Endocrine and Reproductive Systems

Nervous and Endocrine Systems

Urinary and Digestive Systems

Respiratory and Cardiovascular Systems

3. Which of these organs belong to each of the organ systems listed below?

a. Biceps brachii
b. Blood vessels
c. Brain
d. Bronchi
e. Esophagus
f. Femur
g. Hair
h. Heart
i. Kidneys
j. Large intestine
k. Liver
l. Lungs
m. Ovaries

n. Pancreas
o. Pharynx
p. Rectus femoris
q. Sciatic Nerve
r. Skin
s. Spleen
t. Sternum
u. Testes
v. Thyroid
w. Ureters
x. Urinary bladder
y. Uterus
z. Vagina

_____ Integumentary System

_____ Skeletal System

_____ Muscular System

_____ Nervous System

_____ Endocrine System

_____ Cardiovascular System

_____ Lymphatic System

_____ Respiratory System

_____ Digestive System

_____ Urinary System

_____ Reproductive System

CELL ANATOMY AND LIFE CYCLE

Lab 3

Objectives

1. Explain the functions of cell organelles
2. Describe the stages of the cell life-cycle
3. Identify phases of mitosis in microscope slides

Materials

- Models of cells
- Diagrams of cells
- Prepared microscope slides
  - Giant multipolar neurons
  - Small intestine simple columnar epithelium
  - Sperm smear
  - Stages of mitosis in whitefish

INTRODUCTION

This is not a course in cell biology, but a basic understanding of cells is the foundation of study in anatomy and physiology. The **cell** is the smallest living unit in the body. Multiple cells of the same type together form a **tissue.** And **organs** are made up of multiple tissue types. As we study the microscopic anatomy and the physiology of the various organ systems, you will refer again and again to the workings of the cell.

Most cells are subject to wear and aging. If your cells did not replace themselves you would quickly disintegrate! The body's mechanism for replacing lost or damaged cells relies on the cell life cycle.

In this lab you will review the functions of cellular structures, examine various different types of cells, and identify the stages of the cell life cycle.

CONCEPTS

I. Functions of Cell Structures

The boundary of the cell is defined by the **plasma membrane.** Within the plasma membrane are **organelles** that are responsible for cellular operations. Table 3.1 lists cellular structures and their functions that are important for a basic understanding of the cellular basis of anatomy and physiology.

II. Cell Shape and Specializations

There are trillions of cells in the human body. Each cell type is specialized for a specific function and has a unique structure. As you observe three different cell types in this exercise, take note of their specialized structures. Consider the relationship between the cell shape and its unique functions.

Table 3.1
Functions of Cell Structures

◆◆◆

Organelle	Description and Function
Plasma membrane	a double layered membrane of phospholipids that separates the intracellular and extracellular environments; the membrane has many proteins embedded within it that perform various functions
Nucleus	the largest organelle is defined by the nuclear envelope, a phospholipid membrane similar to the plasma membrane; the nucleus contains the chromatin, which is the genetic material
Cytoplasm	the fluid and all organelles outside the nucleus

Cytoplasmic Organelles

a. *mitochondria*—membranous organelles that are instrumental in the synthesis of ATP; contain an inner membrane called the **cristae** that creates a maze-like appearance

b. *ribosome*—non-membranous organelle that provides a site for mRNA translation during protein synthesis

c. *rough endoplasmic reticulum*—membranous organelle dotted with ribosomes that is continuous with the nuclear envelope; proteins are synthesized on ribosomes and packaged in vesicles for transport to the Golgi apparatus; phospholipids are synthesized

d. *smooth endoplasmic reticulum*—membranous organelle synthesizes and metabolizes lipids; involved in drug detoxification; no ribosomes are present

e. *Golgi apparatus*—membranous warehouse that modifies and packages proteins for transport outside of the cell or to lysosomes

f. *lysosome*—membranous sphere that contains digestive enzymes

g. *peroxisome*—membranous sphere that contains detoxifying enzymes

h. *centrioles*—non-membranous, paired organelles composed of microtubules; involved in formation of mitotic spindle in preparation for cell division

i. *microvilli*—finger-like projections of cytoplasm from the plasma membrane that increase absorptive surface area

j. *cilia*—hair-like projections from the cell composed of microtubules that move in unison to create a current across the surface of the cell; instrumental in moving fluids or mucus across the cell surface

k. *flagellum*—tail-like projection composed of microtubules that propels the cell forward; only human cell with a flagellum is the sperm

Nuclear Structures

a. *nuclear envelope*—a double-layered membrane perforated with pores through which substances can move in or out

b. *nucleolus*—non-membranous body composed of rRNA; synthesizes ribosomal subunits

c. *chromatin*—DNA strand combined with histone proteins loosely scattered within the nucleus; DNA is the genetic material

◆ Exercise 3.1 *Examine Three Different Cell Types*

A. Nerve Cells (Photo 366a)

Nerve cells (neurons) have **cell bodies** from which extend long cytoplasmic protrusions. The cell body contains the **nucleus** and is the site of metabolism and protein synthesis. Nerve cells conduct electrical signals along their **cytoplasmic extensions.**

Giant Multipolar Neurons are located within the spinal cord. View a Giant Multipolar Neuron under the microscope with the 10X and 40X objectives. Identify the **nucleus** in the cell body and the **nucleolus** within the nucleus. Observe the **cytoplasmic extensions** from the cell body. Try to identify the small blue "dots" in the cell body and cytoplasmic extensions. These are a form of rough endoplasmic reticulum called Nissl bodies.

Draw your observations here:

B. Simple Columnar Epithelial Cells (Photos 354c and 354d)

Epithelial cells form membranous sheets that line cavities or cover organs. Epithelial cells have a variety of functions that depend on their structure and location. The epithelium lining the small intestine is involved with absorption of food subunits into the blood.

Obtain a microscope slide with simple columnar epithelium from the small intestine and view under the 40X objective. This slide will have several tissue types. The simple columnar epithelium is the lining of the lumen (hole) in the small intestine. It may be stained a dark pink or purple. Note the **nucleus** and the **microvilli** on the free edge of the cells.

Draw your observations here:

C. Sperm Cells (Photos 397c and 397d)

Sperm cells are the only human cells that have a flagellum to propel them through a fluid environment. These cells deliver paternal genetic material to the maternal egg. An egg fertilized by a sperm will begin to develop into an embryo.

View a microscope slide with a sperm smear under 40X and using the oil immersion 100X lens. This cell type has very little cytoplasm. The **nucleus** is contained under a rigid cap at the head of the sperm. The **mitochondria** are located at the base of the **flagellum** where it joins the head.

Draw your observations here:

▶ How is the shape of the nerve cell specially suited to the function of that type of cell?

▶ How is the shape of the simple columnar epithelial cells suited to their function as they line the digestive tract?

▶ How is the structure of the sperm cell specially suited to the function of that type of cell?

◆◆◆

III. Cell Life Cycle

Cells undergo a life cycle which has two main phases (Figure 3.1). **Interphase** is the longest phase of the life cycle. During this stage the cell grows and synthesizes proteins. As the cell prepares to enter the cell division stage of the life cycle, the DNA replicates so that there are two copies of the complete genome within the cell. Cell division is divided into two stages: genetic material is evenly parceled out into two packages during **mitosis,** and the cellular cytoplasm splits into two daughter cells during **cytokinesis.** Mitosis is further subdivided into four phases. The major events of each phase are listed in Table 3.2.

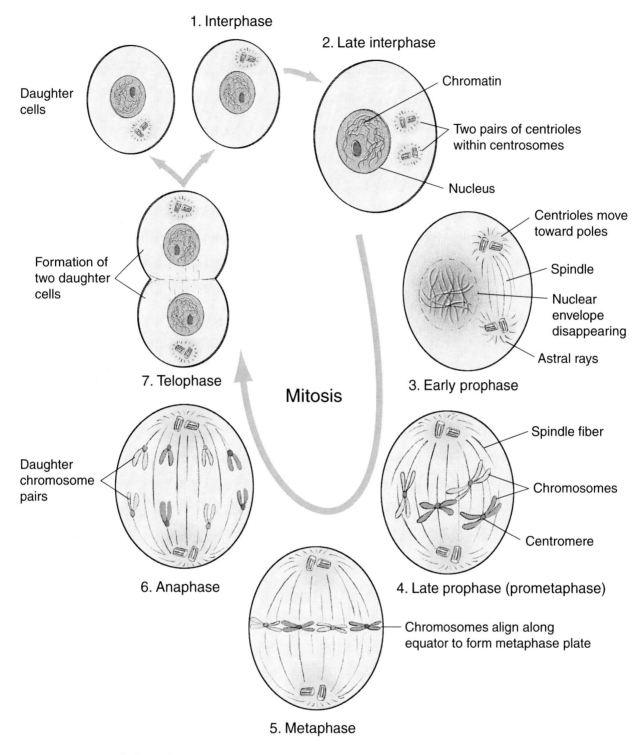

FIGURE 3.1 ◆ Cell Life Cycle

Table 3.2
Phases of the Cell Life Cycle

◆◆◆

Interphase

- normal metabolic activities
- cellular growth
- genetic material in the form of **chromatin;** not visible in the light microscope
- centrioles replicate
- DNA replicates
- Protein synthesis to support cell division

Mitosis

Prophase

- **Asters** extend from the centrioles
- Chromatin coils and condenses into **chromosomes;** visible in the light microscope
- Chromosomes are made of two sister **chromatids** joined together by a **centromere**
- Centriole pairs separate to opposite poles of the cell
- Microtubules extend from centrioles to form the **mitotic spindle**
- Nuclear envelope fragments and disappears
- Nucleolus disappears
- Mitotic spindles occupy the center of the cell
- Special microtubules called **kinetochore** microtubules attach to centromeres on chromosomes

Metaphase

- The plane in the center of the cell is called the **metaphase plate**
- Chromosomes align their centromeres at the center of the mitotic spindle at the metaphase plate

Anaphase

- Centromeres split and the two sister chromatids from each chromosome are now each called chromosomes themselves
- Cell elongates
- Kinetochore microtubules shorten and pull chromosomes to opposite poles of cell
- Chromosomes shaped like Vs
- **Cleavage furrow** begins to form

Telophase

- Chromosome movement stops
- Chromosomes uncoil into chromatin
- Nuclear envelope reforms from rough ER around chromatin
- Nucleolus reforms
- Mitotic spindle breaks down
- Cell is briefly binucleate
- Cleavage furrow deepens

Cytokinesis

- Contractile ring of microfilaments forms at the cleavage furrow and squeezes the cell into two daughter cells

◆ Exercise 3.2 *Examine Whitefish Mitosis Microscope Slides*

Obtain a Whitefish Mitosis microscope slide and examine it under 40X power under the microscope. Refer to Photos 353a–353e to identify cells in interphase and the stages of mitosis.

For each of the stages of the cell life cycle list the distinguishing features that will help you identify it. Draw each stage below.

Interphase

Distinguishing features:

Prophase

Distinguishing features:

Metaphase

Distinguishing features:

Anaphase

Distinguishing features:

Telophase

Distinguishing features:

Name _____

Lab 3: Cell Anatomy and Life Cycle

◆ Practice

1. Choose the appropriate cell structure for each of the functions listed below.

   a. centrioles
   b. chromatin
   c. cilia
   d. cytoplasm
   e. flagellum
   f. Golgi apparatus
   g. lysosomes
   h. microvilli
   i. mitochondria
   j. nuclear envelope
   k. nucleolus
   l. peroxisomes
   m. ribosomes
   n. rough endoplasmic reticulum
   o. smooth endoplasmic reticulum

   _____ hair-like structure that propels mucus or fluid across cell surface

   _____ membranous sac that contains digestive enzymes

   _____ the fluid and all organelles outside of the nucleus

   _____ membrane with pores that contains the genetic material and the organelle that makes ribosomal subunits

   _____ tubular membranes involved in the synthesis of lipids and in drug detoxification

   _____ the primary site for synthesis of ATP in the cell

   _____ membranous structure that packages proteins for export or for use within the cell

   _____ pairs of structures that use microtubules to create the mitotic spindle

   _____ extensions of the cytoplasm that increase surface area for absorption of extracellular material

   _____ non-membranous structure with a binding site for mRNA

2. Starting with the rough ER, follow the path of a protein that will be secreted from the cell as a product.

3. What is the difference between chromatin and chromosomes? In which cell life cycle stage(s) would you find each?

4. What is the significance of the fact that the membranes of all membranous organelles (except mitochondria) have the same composition?

5. During which stage of the cell life cycle would you see the following events?
   a. Interphase
   b. Prophase
   c. Metaphase
   d. Anaphase
   e. Telophase
   f. Cytokinesis

   _____ chromosomes uncoil into chromatin

   _____ sister chromatids split into V-shaped chromosomes

   _____ centrioles move to opposite poles of the cell

   _____ cell growth and metabolism occurs

   _____ cleavage furrow constricts cytoplasm to form two daughter cells

   _____ chromosomes align at the center of the mitotic spindle

   _____ nuclear envelope fragments and disappears

6. Which stage of the cell life cycle would you be viewing if you observe the following features under the microscope?

   Elongated cell with two nuclei

   Indistinct genetic material and no mitotic spindle

V-shaped dark chromosomes with cleavage furrow

Cleavage furrow with dark chromosomes in two distinct clusters

Dark chromosomes aligned across the center of the cell

PASSIVE TRANSPORT MECHANISMS THROUGH PLASMA MEMBRANES

Lab 4

Objectives

1. Define **Brownian Motion** and **semipermeable membrane**
2. Explain the passive transport processes of **diffusion** and **osmosis**
3. Predict which way substances will move through a semipermeable membrane given information about concentration and osmolarity
4. Describe **isotonic, hypotonic** and **hypertonic** solutions and their effect on the movement of water through a membrane

Materials

- Demonstration 4.1:
  - Insoluble dye (Carmine powder) for Brownian Motion demo
  - Microscope slides
  - Coverslips
  - Small pin or dissecting needle
  - Water
  - Dropper
- Exercise 4.1:
  - Petri plates with non-nutrient agar (1.5% agar)
  - Potassium permanganate crystals
  - Forceps
- Demonstration 4.2:
  - Red onion
  - Microscope slides
  - Coverslips
  - Forceps
  - Scalpels
  - Dropper bottles containing:
    - distilled water
    - 1% NaCl
    - 10% NaCl

- Demonstration 4.3:
  - Thistle tube
    - Thistle tube bottom covered with dialysis membrane, tightly wrapped with rubber band
    - Thistle tube bulb filled with 40% sucrose solution, colored with food color
  - Holder for thistle tube (thermometer holder)
  - Ring stand
  - Thistle tube bulb immersed in beaker of distilled water
- Exercise 4.2:
  - Balances
  - Solutions (for six lab groups):
    - about 250 ml 10% NaCl
    - about 200 ml 0.1% boiled starch
    - about 350 ml 40% glucose (in 50-ml Erlenmeyers)
    - about 150 ml/beaker x 6 beakers of 40% glucose
  - For Each Lab Group:
    - hot plate
    - four 250-ml beakers
    - one 400-ml beaker, hot water bath w/boiling chips
    - small funnel
    - wax pencil
    - test tube rack with test tubes
    - spot plate
    - 40% glucose (in small Erlenmeyer and in one of the 250-ml beakers)
    - 0.1% starch (in small Erlenmeyer)
    - 10% NaCl (in small Erlenmeyer)
    - test tube holders (4)
    - transfer pipets
    - Benedict's solution
    - Silver nitrate ($AgNO_3$) solution
    - Lugol's iodine solution
    - Four pieces of dialysis tubing cut into about 4-inch lengths, soak in tap water

INTRODUCTION

In order for you to survive you must carefully regulate what substances enter your body. If the wrong substances enter your body, or if too much of some substances enter your body, you will experience homeostatic imbalances that could lead to death. In the same way, if too much of a substance, such as blood, leaves your body, this could be detrimental to your health. Individual cells also rely on regulatory mechanisms that control what substances enter and exit. The health of your body cells, and thus your own health, depends on the cell's ability to maintain an internal environment conducive to metabolism.

The human body is mostly (60–80%) water, which contains gases and dissolved particles, or **solutes,** that are necessary for survival. Solutes may include nutrients, waste particles and ions. Plasma membranes define the borders of cells and create internal (**intracellular**) and external (**extracellular**) spaces. Water, gases and solutes must cross the plasma membrane to move from one space to the other. There are several mechanisms that control the transport of gases, solutes and water across the plasma membrane.

Transport mechanisms are either active or passive. **Active** processes require energy to move solutes through the plasma membrane in the direction that, without energy, the solutes could not otherwise move. Transporting a barge upriver is an example of an active process. Without energy to fuel the engines that enable the barge to move against the current, the boat will simply float downstream. **Passive** transport mechanisms, on the other hand, happen all the time without the need for energy. When the barge floats downstream it is moving passively. Water, solutes and gases all move passively across membranes in response to concentration or pressure gradients.

For any molecule to pass through the lipid bilayer of the plasma membrane, the membrane must be *permeable* to that substance. There are several physical ways a molecule can cross the membrane. It may be small enough to pass through the naturally existing pores within the membrane, or may be able to dissolve across the lipid bilayer. Oftentimes there are protein carriers or channels that allow a particular molecule to pass. Any cell membrane will allow some substances to cross while preventing others. This property is called **semipermeable.**

The exercises in this lab illustrate passive transport mechanisms across semipermeable membranes.

CONCEPTS

I. Brownian Motion

Fluid molecules are constantly moving in a random, irregular manner. This kinetic energy is transferred to solutes dissolved within the fluid as water molecules jostle and bounce the solutes from all directions. Thus, even though the solutes have no motility of their own, they appear to be moving and vibrating because of the movement of the fluid molecules around them. This solute movement is called **Brownian Motion** and is responsible for the dispersement of a solute dissolved in fluid.

◆ **Demonstration 4.1**   *Observing Brownian Motion*

Your instructor will set up a wet mount microscope slide that contains dye crystals suspended in a drop of water. The dye crystals should appear to be vibrating and shaking, sometimes moving from place to place. They are moving because they are colliding with smaller water molecules that cannot be seen.

II. Diffusion and Osmosis

A. Simple Diffusion

Simple diffusion is the passive movement of a solute from an environment of high concentration to an area of low concentration. This difference in concentration of a particular solute is called a **concentration gradient.**

◆ **Exercise 4.1** *Visualize a Concentration Gradient*

Agar is a semi-solid medium that has a higher kinetic energy within its molecules than a solid substance. In this experiment you will observe the diffusion of dye molecules from a highly concentrated area to an area of low concentration.

1. Using forceps, obtain a single large crystal of potassium permanganate from the labeled container.
2. Place one crystal on the agar surface of a prepared petri dish. Do not pour crystals directly from the potassium permanganate container onto the agar. If you get too many crystals you will not be able to effectively see the diffusion of dye molecules away from the concentrated crystal into the agar.
3. At 15 minute intervals observe the growing halo of dye within the agar.

   ▶ How does the color intensity of the farthest diameter compare to that of the crystal?

4. Explain your observations in terms of diffusion:

B. Osmosis

Osmosis is the passive movement of water across a semipermeable membrane. To understand osmosis you first must understand the concept of osmolarity. The **osmolarity** of a fluid is the total number of all types of solutes dissolved within a given volume of the fluid. To put it another way, osmolarity is the concentration of all solutes in a fluid. Water moves as a result of osmotic pressure across a membrane, moving from the environment with the lower osmolarity toward the area with higher osmolarity. In effect, water moves in the direction that will dilute the higher solute concentration. We can describe the fluid environment in which a cell is bathed as either hypotonic, hypertonic, or isotonic.

A **hypotonic** solution has a lower osmolarity than the intracellular fluid (*hypo*- below). Water will passively move from the lower osmolarity extracellular environment into the cell, and will eventually cause the cell to burst.

A **hypertonic** solution has a higher osmolarity than the intracellular fluid (*hyper*- above).

   ▶ Can you explain why a cell in a hypertonic solution will shrink?

Cells within the human body exist in an **isotonic** (*iso*- same) environment, where intracellular and extracellular osmolarities are the same.

◆ **Demonstration 4.2** *Osmosis in Onion Cells*

Your instructor will demonstrate the effects of tonicity on onion cells. Plant cells, unlike animal cells, have a rigid cell wall that prevents a cell from bursting upon the rapid influx of fluid. Three thin slices of red onion are placed on a microscope slide. Each slice is bathed in either 1.0% NaCl, 10% NaCl, or distilled water.

Predict the tonicity of each solution by matching the solutions with the corresponding tonicity:

1.0 % NaCl	hypotonic
10% NaCl	hypertonic
distilled water	isotonic

Record the effects of each solution on the onion cells in Table 4.1.

Table 4.1
Results for Demonstration 4.2

◆◆◆

Solution	Effect on Onion Cells	Tonicity of Solution

◆ **Demonstration 4.3** *Osmosis Through Nonliving Membrane in a Thistle Tube*

To illustrate the movement of water across a semipermeable membrane, a thistle tube osmometer has been prepared for you. A thistle tube is a glass container with a bulb-like base that is open at the bottom. A long tube extends upward from the bulb and is open at the top. In this demonstration, the bottom of the bulb is covered tightly by a semipermeable dialysis membrane. The bulb is filled to the bottom of the long tube with a 40% sucrose solution. Coloring has been added for better visualization. The membrane bound bulb was immersed in a beaker of distilled water at the beginning of your class.

1. Check the tube at 15 min. intervals and note the height of the water in the thistle tube.

2. Explain your observations by answering the following questions:

   ▶ At what relative height did the solution stop rising?

   ▶ Why did the solution rise inside the thistle tube?

   ▶ The dialysis membrane is impermeable to sucrose, a sweet monosaccharide. If you had tasted the beaker water at the end of the experiment, what would you have noticed? How would you explain this?

◆ **Exercise 4.2** *Predict Diffusion and Osmosis in an Artificial Cell*

Keep in mind that both simple diffusion of a solute and osmosis can occur simultaneously through the same membrane. However, *net* osmosis (a net movement of water from one side of the membrane to the other) will only occur if the membrane is *impermeable* to the solute that is present. Explain this fact when you record your results in this exercise.

This experiment will take an entire class period. You will be testing for the movement of water and/or solute across a nonliving membrane in response to changes in the relative concentrations of solute and solvent. The sac represents a cell, and the beaker represents the environment in which the cell exists.

1. Work in groups of four. Obtain for your lab group the equipment and solutions listed at the beginning of this lab under *Materials* for Exercise 4.2.
2. Obtain four dialysis sacs from a beaker of water. They have been soaking in order to increase flexibility and to allow you to open one end. Also cut eight strands of string, each about six inches long.

3. Label four beakers #1, #2, #3, #4 with the wax pencil. Fill beakers #1, 3, and 4 half full with tap water. Fill beaker #2 half full with a 40% glucose solution from your table.

4. Tie off one end of each of the dialysis tubes by twisting the end, folding the end back on itself and tightly tying the string so that the end remains folded. A leaky sac will skew the results of your experiment.

5. Using the funnel, half fill each sac with the solution specified below. Be sure to rinse the funnel between solutions. Making sure there is not a lot of air in the sac, tie off the open end as you did in step 4. Cut off the ends of the string.
   • Sac 1: 40% glucose solution
   • Sac 2: 40% glucose solution
   • Sac 3: NaCl solution
   • Sac 4: Boiled starch solution

6. Carefully blot dry each sac and weigh it. Make sure to tare the scale before weighing each sac. Record the weights of each sac in grams (g) in Table 4.2.

7. Place each sac in its corresponding beaker. Make sure the sac is completely submersed in the solution. Leave the sacs immersed in the solutions for at least 1 hour. The longer you leave the sacs, the clearer will be your results.

8. Predict whether osmosis or diffusion or both will occur through the dialysis membranes for each of the sacs and record your predictions in Table 4.2. Include your prediction for which direction osmosis or diffusion will occur (into or out of the sac).

9. About 40 minutes prior to the end of class, fill the fifth beaker to about one inch from the top with water and start it boiling on the hot plate. Retrieve Sac 1 from Beaker 1 and **gently** blot it dry without squeezing it. Weigh Sac 1 and record the weight in Table 4.2.

10. Place 2 ml of Benedict's solution in each of two test tubes labeled "B" and "S." Put 2 ml of fluid from Beaker 1 in the "B" test tube and 2 ml of fluid from Sac 1 in the "S" tube. Use two test tube holders per tube to suspend each tube in the boiling water. Boil the tubes for two minutes. The formation of a precipitate, usually white in color, indicates a positive test for the presence of glucose in the solution. Record your results in the chart.

11. Blot gently and weigh Sac 2. Record your results.

12. Blot gently and weigh Sac 3. Add 4 ml of fluid from Beaker 3 to a clean test tube. Add a drop of silver nitrate. The formation of a white precipitate indicates a positive test for the formation of silver chloride, which is formed when silver nitrate reacts with sodium chloride. Record your results.

13. Blot gently and weigh Sac 4. Remove 4 ml of fluid from Beaker 4 to a clean test tube and add a couple of drops of Lugol's iodine. The formation of a black color indicates a positive test for the presence of starch. Record your results.

14. Clean-up guidelines:
    a. Throw away sacs, used plastic pipets and Petri plates into the trash.
    b. Dump test tube fluids in the sink and place test tubes in a designated wash tub.
    c. Save the glucose from beaker #2 for future use by covering with plastic wrap.
    d. Dump water from beakers #1, 3 and 4 in the sink. Do not attempt to dump the hot water from the hot water bath unless it is cooled.

Table 4.2
Results for Exercise 4.2

◆◆◆

#: Beaker Contents: Sac Contents:	1 Water 40% glucose	2 40 % glucose 40 % glucose	3 Water NaCl	4 Water Boiled starch
a. Weight at start:				
b. Prediction:				
c. Weight at end:				
d. % Change in weight: =(c–a)×100/a				
e. Test results: (+/–)	Sac fluid: Beaker fluid:		Beaker fluid:	Beaker fluid:
f. Did osmosis occur? Which direction?				
g. Did diffusion occur? Which direction?				

Name _____

Lab 4: Passive Transport Across Plasma Membranes

◇ Practice

1. A group of students made three sacs out of dialysis tubing and filled them: Sac A = 20% glucose; Sac B = 10% NaCl; Sac C = starch solution

   Each sac was placed in a beaker of distilled water and allowed to sit for an hour, then the beaker water was tested for certain substances. Answer the following questions based on the experiment results:

   a. The beaker water with Sac A tested positive with Benedict's solution. What substance moved from the sac into the beaker?

   b. Name the process by which the substance moved out of Sac A.

   c. The beaker water with Sac B tested positive for silver chloride, and the sac was heavier than at the beginning of the experiment. What substance moved into the sac and what moved out?

   d. Name the processes by which each of the substances from c above moved through the membrane.

   e. The starch test is *negative* in beaker C and the sac is larger (swollen) than it was at the beginning of the experiment. What substance moved into the sac?

   f. Name the process by which the substance moved into Sac C.

   g. Explain why the starch did not move into the beaker from the sac.

2. A dialysis sac containing 20% glucose is immersed in a beaker of unknown glucose concentration and allowed to sit for two hours. The beginning weight of the sac was 20g. The end weight of the sac is 16g. Based on this information:

   a. What is the % weight loss of the dialysis sac?

   b. The fluid in the beaker is _____ to the fluid in the sac (hypertonic, isotonic, hypotonic).

   c. What substances moved through the dialysis tubing and in which direction did they move (in or out of sac)?

   d. If the sac contained a 10% glucose solution, would the process observed occur faster or slower? Why?

3. The motion of non-living solutes in water is termed _____.

4. Define *semipermeable membrane*.

5. What is the difference between a *concentration gradient* and an *osmotic gradient?*

6. All of the exercises and demonstrations in this lab are examples of passive transport mechanisms. What is the difference between passive and active transport mechanisms?

7. Use the concept of *tonicity* to explain why an intravenous fluid of 0.9% saline is preferable to distilled water. What would happen to blood cells if distilled water were delivered intravenously?

HISTOLOGY

Objectives

1. Describe the location and functions of the various tissue types
2. Use the microscope to identify the various tissue types

Materials

- Microscope slides of epithelial and connective tissue types

INTRODUCTION

We began this course by studying the smallest basic living unit within the human body—the cell. You will have learned that when individual cells of similar structure and function are grouped together they form a tissue. Multiple tissues that function together toward a common purpose form an organ, and related organs work collectively in an organ system.

Histology is the study of tissues. Each tissue type has specialized functions that contribute directly to the physiological processes of the organs in which it is located. Therefore, before we study individual organ systems it is prudent to become familiar with the different tissue types and subtypes.

There are four main tissue types with specialized characteristics and functions: **epithelial, connective, muscle and nervous.** Each of these types can be further subdivided into specific subtypes that are ideally suited for their specific locations and purpose. In this lab you will make a thorough examination of epithelial and connective tissue subtypes. You will have an opportunity for observation of muscle and nervous tissues in later labs.

CONCEPTS

I. Epithelial Tissue

Epithelial tissue, called epithelium, typically forms sheets or membranes that line organs or cavities. Functions of epithelia include protection of underlying tissues, secretion of products, excretion of wastes, filtration and absorption. Epithelial membranes are attached to another tissue or organ on the **basal** surface, but the **apical** surface is unattached to any other tissue. A **basement membrane** adheres the basal surface to the underlying tissue. Epithelia have no blood supply (**avascular**), and rely upon diffusion of nutrients and oxygen from adjacent vascular tissues. They do have a nerve supply (**innervated**). Epithelia are classified according to two characteristics: the number of layers of cells and the shape of the cells.

A single layer of cells is termed a **simple** epithelium, while a membrane with multiple epithelial cell layers is called a **stratified** epithelium. **Pseudostratified** refers to an epithelium in which all cells are in contact with the basal surface, but are various heights. Thus a pseudostratified epithelium has nuclei at several heights in the tissue.

Epithelial cell shapes include squamous, cuboidal and columnar. **Squamous** cells have very little cytoplasm and appear squashed. The nuclei within squamous cells are flattened. **Cuboidal** cells are shaped like a cube, and have spherical nuclei situated in the center of the cell. **Columnar** cells are elongated and have oval nuclei positioned near the basal surface.

Names of epithelia include both the number of layers and the shape of the cells. Thus, simple squamous epithelium is a single layer of squamous cells, but stratified cuboidal is multiple layers of cuboidal cells. Stratified squamous epithelium actually has cuboidal cells at the basal surface that become increasingly flattened toward the apical surface. Various epithelia may also include organelles such as cilia for the movement of mucus, or microvilli for the absorption of nutrients.

Glandular epithelia do not form membranes, but instead form glands that secrete products into their surroundings. **Exocrine** glands secrete products into a duct formed by epithelial tissue (usually simple or stratified cuboidal epithelia). The duct delivers the product, such as saliva or digestive enzymes, to a specific location. **Endocrine** gland products are secreted directly into the extracellular fluid without ducts, and are transported to all body regions by the blood and lymph.

Table 5.1
Epithelial Tissue Locations and Functions

◆◆◆

Tissue	Locations	Functions
Simple Squamous	air sacs of lungs, serous membranes, inner layer of blood vessel wall	diffusion of materials, secretion of fluids, filtration
Simple Cuboidal	walls of some ducts and small glands	secretion of a product, absorption of materials
Simple Columnar		
Non-ciliated	line digestive tract	absorption, secretion of mucus and other substances
Ciliated	line lower respiratory tract and uterine tubes	secretion and movement of mucus, movement of female oocyte (egg)
Pseudostratified Columnar		
Ciliated	line the trachea (windpipe) and upper respiratory tract	secretion and movement of mucus
Stratified Squamous		
Keratinized	skin epidermis	protection of underlying tissues
Non-keratinized	lines the esophagus, mouth, vagina	protection of underlying tissues

◆ Exercise 5.1 *Identify Epithelial Tissue Subtypes*

As you examine the histological slides under the microscope, relate the specific structures and characteristics of the tissue to the functions associated with its location. For example, for what purpose would intestinal epithelial cells have microvilli? (hint: What is the function of microvilli?) Use the following procedure to examine the listed tissue types:

1. Remember that each tissue type is usually closely associated with other tissue types. Therefore, you will probably need to first identify the location of the specific tissue type you are studying on the microscope slide. Use the photo atlas pictures to help you.
2. Begin your microscopic examination of the tissue slide with the lowest power lens. This larger field of view will enable you to see the relationship of the different tissue types that are present in the specimen. Center the region of the slide that has the tissue type you are observing.
3. Use the 10X and 40X objectives for a more detailed analysis of the tissue features.
4. Use colored pencils to sketch and label the tissue you are viewing. This process uses a different part of your brain to learn and memorize tissue features than merely using your eyes.
5. Consider the plane through which you are viewing the tissue. It may be longitudinal or cross sectional, or even an oblique angle. The plane will affect the appearance of the tissue.

A. Simple Squamous Epithelium (Photo 354a)

    1. Draw your observations here:

    2. Location:

    3. Function:

B. Simple Cuboidal Epithelium (Photo 354b)

    1. Draw your observations here:

    2. Location:

    3. Function:

C. Simple Columnar Epithelium (Photos 354c, 354d)

   1. Draw your observations here:

   2. Location:

   3. Function:

D. Ciliated Pseudostratified Columnar Epithelium (Photo 354e)

   1. Draw your observations here:

   2. Location:

   3. Function:

E. Stratified Squamous Epithelium (Photo 354f)

   1. Draw your observations here:

   2. Location:

   3. Function:

◆◆◆

II. Connective Tissue

Connective tissues have a variety of consistencies and appearances. They are all classified together due to their common embryonic origin: all connective tissues arise from the embryonic tissue **mesenchyme.** Four types of relatively undifferentiated cells secrete the four major types of connective tissue. **Fibroblast** cells secrete the various types of **connective tissue proper, osteoblasts** secrete **bone, chondroblasts** secrete **cartilage** and **hemocytoblasts** give rise to **blood cells.** The suffix *blast* indicates the cells' ability to form additional types of tissue or cells. Bone, cartilage and blood are specialized connective tissue types.

As the blast cells mature they become more differentiated and serve to maintain the health of the connective tissues with which they are associated. The suffix of a mature cell is *cyte.* Thus, **fibrocytes** maintain connective tissue proper, **osteocytes** are found in bone, and **chondrocytes** are in cartilage. Hemocytoblasts are stem cells that produce differentiated daughter cells such as erythro*cytes* (red blood cells), leuko*cytes* (white blood cells) and thrombo*cytes* (platelets).

Blast cells secrete the **extracellular matrix,** or non-living portion, of connective tissue. The extracellular matrix includes **fibers** and **ground substance.** There are three types of fibers secreted by blast cells: collagen, elastic and reticular. **Collagen fibers** are thick and resist tension. They provide strength to connective tissues in which they are located. **Elastic fibers,** like their name implies, give the property of elasticity or stretch to the tissue. **Reticular fibers** form branching networks and serve to house blood cells. Not all connective tissues have all three types of fibers. The ground substance of connective tissue is material that contains the fibers and surrounds the cells. It can be fluid, as in blood, or rigid, as in bone.

In general, the functions of connective tissue are to connect different tissues together, to protect other tissues, to insulate, and to transport substances within the body. Cartilage is avascular, and other connective tissues have varying degrees of vascularity.

Table 5.2
Connective Tissue Properties

◆◆◆

Tissue	Description	Locations	Functions
Connective Tissue Proper			
Areolar (Loose)	collagen, elastic and reticular fibers in a gelatinous matrix; cell types include fibroblasts, macrophages, mast cells, white blood cells	under epithelia, surrounds some organs and blood vessels	cushion organs, house macrophages to defend against infection, contain water
Dense Regular	many parallel collagen fibers provide great strength in two directions; a few elastic fibers are present; cell type is fibroblasts; vascular	tendons, ligaments, aponeuroses	connect bone to muscle or bone, withstand unidirectional tension
Dense Irregular	many collagen fibers arranged in all orientations provide great strength in all directions; a few elastic fibers are present; cell type is fibroblasts; vascular	skin dermis, digestive tract submucosa, fibrous capsules around joints and organs	withstand multidirectional tension, protection of joints and organs
Adipose	collagen, elastic and reticular fibers in a sparse gelatinous matrix; adipocytes have large spaces within thin cytoplasm that contain fat; highly vascular	under skin (hypodermis), around kidneys, in abdomen, breasts	protection from physical trauma and temperature extremes, energy storage and source
Reticular	networks of reticular fibers secreted by fibroblasts house red and white blood cells; vascular	lymph nodes, spleen	houses red and white blood cells
Specialized Connective Tissues			
Bone	rigid matrix contains collagen fibers and calcium salts; osteoblasts and osteocytes are present; highly vascular	skeleton	provide support, protection for vital organs, mineral storage, blood cell formation
Cartilage	collagen fibers are present but not visible in the microscope; matrix is firm but resilient; cell types are chondroblasts and chondrocytes; avascular		
• Hyaline		trachea, ends of long bones, ribs, nose	provide support with some resiliency
• Elastic		external ear, epiglottis	provide structure with flexibility
• Fibrocartilage		intervertebral discs, pubic symphysis, menisci of knee (and other) joints	firm, shock-absorbing support with great strength
Blood	fluid matrix is called plasma, contains no fibers; erythrocytes, leukocytes and platelets are present	within blood vessels and heart	carry respiratory gases, nutrients, hormones, waste products; white blood cells protect from infection

◆ **Exercise 5.2** *Identify Connective Tissue Subtypes*

Use the guidelines for tissue identification in Exercise 5.1 to help you identify the connective tissue subtypes.

A. Connective Tissue Proper

1. Areolar Connective Tissue (Photo 355a)

    a. Draw your observations here:

    b. Location:

    c. Function:

2. Dense Regular Connective Tissue (Photo 355b)

    a. Draw your observations here:

    b. Location:

    c. Function:

3. **Dense Irregular Connective Tissue: (Photo 355c)**
   a. Draw your observations here:

   b. Location:

   c. Function:

4. **Adipose Tissue (Photo 355d)**
   a. Draw your observations here:

   b. Location:

   c. Function:

5. **Reticular Connective Tissue (Photo 355e)**
   a. Draw your observations here:

   b. Location:

   c. Function:

B. **Specialized Connective Tissues**
   1. **Bone (Compact bone) (Photo 360a, 360b)**
      a. Draw your observations here:

      b. Location:

      c. Function:

2. **Cartilage (Hyaline) (Photo 359a)**
   a. Draw your observations here:

   b. Location:

   c. Function:

3. **Blood (Photos 377a–381c)**
   a. Draw your observations here:

   b. Location:

   c. Function:

Name _____

Lab 5: Histology

◆ **Practice**

1. Answer the following questions about epithelial tissue:

    a. Describe the general characteristics of epithelial tissue.

    b. On what bases are epithelial tissues classified?

    c. What are the major functions of epithelia in the body? (Give examples.)

    d. Where is ciliated epithelium found? What role does it play?

    e. How do the endocrine and exocrine glands differ in structure and function?

2. Answer the following questions about connective tissue:

    a. What are the general characteristics of connective tissue?

b. What functions are performed by connective tissue?

c. How are the functions of connective tissue reflected in its structure?

3. Write in the name of the specific tissue type for each of the descriptions below.

_____ found in the peritoneum, secretes peritoneal fluid

_____ ciliated variety is found in the *lower* respiratory tract

_____ acted upon by muscles to allow movement

_____ houses blood cells within the lymph nodes

_____ surrounds many organs, contains water

_____ absorbs materials from inside a duct

_____ absorbs digested food subunits from the digestive tract

_____ source of concentrated fuel, protects vital organs

_____ propels mucus superiorly in the trachea

_____ withstands tension from a muscle pulling on a bone

THE INTEGUMENTARY SYSTEM

Lab 6

Objectives

1. Identify the structure of the skin and its accessory organs on a skin model
2. Explain the functions of skin and skin appendages
3. Observe the differences between thin and thick skin on microscope slides
4. Observe the differences between pigmented and nonpigmented skin in microscope slides
5. Describe the significance of skin color changes

Materials

- Model of skin
- Microscope slides
  - Thin skin
  - Thick skin
  - Skin with hair follicles
  - Pigmented vs. non-pigmented skin

INTRODUCTION

At last we turn our studies to organ systems! The integumentary system includes the skin and its accessory structures– sweat glands, oil glands, hair and nails. The skin not only defines the boundary between your internal and external environments, it also has protective and regulatory functions. Underlying tissues are protected from physical, chemical or temperature induced trauma. The internal body temperature is also regulated in part by structures within the skin. Cutaneous blood vessels dilate to allow the hot blood to come closer to the surface of the body and radiate heat to the cooler environment. This, coupled with evaporation of sweat from sweat glands (sudoriferous glands), cools the body. Alternatively, cutaneous blood vessels can constrict to keep the warmth of the blood within the core of the body and maintain warmth in a cooler environment. Metabolic functions include the utilization of sunlight to convert a type of cholesterol into a vitamin D precursor.

CONCEPTS

Table 6.1
Layers of the Skin

◆◆◆

Epidermis—Keratinized Stratified Squamous Tissue (See Figures 6.1 and 6.2.)

The epidermis is avascular, receiving its nutrients from capillaries in the underlying dermis. It is composed of four layers in thin skin, with a fifth layer in the thick skin of the palms and soles.

- **Stratum Basale**—basal layer
  - mitotically active **keratinocytes**
  - **melanocytes** secrete the pigment melanin
  - **Merkel cells** are receptors that convey information about touch to sensory receptors in the dermis
- **Stratum Spinosum**—spiny layer
  - Keratinocytes become spiny as they begin to shrink
  - **Langerhans' cells** are macrophages that protect against infection
- **Stratum Granulosum**—granular layer
  - Keratinocytes flatten as their nuclei and organelles disintegrate
  - Granules are released that form keratin
- **Stratum Lucidum**—clear layer (thick skin only)
  - Keratinocytes are dead, clear and flattened
- **Stratum Corneum**—cornified layer
  - Keratinocytes are dead, flat; keratinocytes slough off; keratin creates a tough, waterproof, protective surface

Dermis (See Figure 6.2)

The dermis is composed of two distinct layers that house fibroblasts, macrophages and a few mast cells and white blood cells.

- **Papillary layer**—20% of the dermis
  - Areolar connective tissue
  - Dermal papillae are small bumps that protrude superficially into the epidermis;
    - capillaries within some dermal papillae supply the overlying strata with nutrients
    - free nerve endings in some of the papillae act as pain receptors
    - Meissner's corpuscles in some papillae are touch receptors
  - Dermal ridges in the palms and soles push epidermal layers into epidermal ridges, which allow for increased friction and grip, and which make "fingerprints"
- **Reticular Layer**—80% of the dermis
  - Dense irregular connective tissue
  - Contains collagen fibers, which provide strength

Hypodermis

The hypodermis is not actually part of the skin, but is the layer of adipose tissue deep to the dermis.

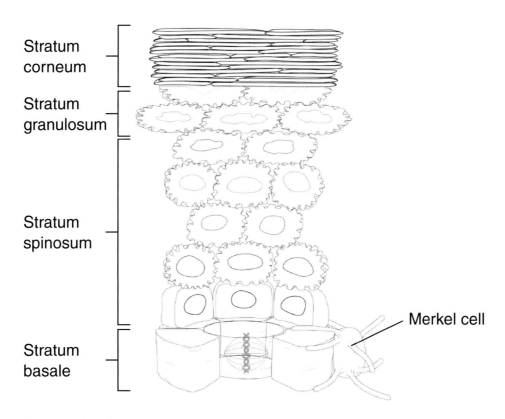

Stratum corneum

Stratum granulosum

Stratum spinosum

Stratum basale

Merkel cell

FIGURE 6.1 ◆ Layers of Epidermis

Table 6.2
Appendages of the Skin

◆◆◆

(See Figure 6.2.)
- **Sweat (Sudoriferous) Glands**
  - **Eccrine** glands—all over entire body, particularly on palms, soles and forehead; coiled tube in dermis secretes sweat into pore in the epidermis
  - **Apocrine** glands—located in axillary and anogenital regions; secrete a fatty sweat with proteins that, when digested by bacteria, create an odor; not significant in cooling the body
  - **Ceruminous** glands—located in external ear canal; secrete cerumin, (ear wax)
  - **Mammary** glands—located in the breasts; secrete milk
- **Sebaceous Glands**—located all over the body except on the palms and soles; secrete sebum, which is oil, into hair follicles; softens and lubricates hair and skin
- **Hair** and **Hair Follicles**
  - **Hair follicle**—an invagination of the epidermis down into the dermis
  - **Hair bulb**—the rounded, expanded portion at the deep end of the follicle
  - **Hair papilla**—contains capillaries and protrudes up into the root of the hair from the hair bulb to supply the hair with nutrients
  - **Hair**—contained within the follicle; that portion superficial to the epidermis is the shaft, and deep to the surface of the skin is the root
  - **Arrector pili muscle**—attaches the follicle to the stratum basale and causes the hair to stand up (piloerection)
  - **Root hair plexus**—a sensory nerve that surrounds the hair bulb and acts as a sensory receptor
- **Nails**—the region of the epidermis on the dorsal surface of fingers and toes that resembles protective scales

Table 6.3
Skin Color and Clinical Significance

◆◆◆

Bronzing—possible indication of Addison's disease, hyposecretion of the adrenal medulla endocrine gland

Bruising—bleeding from blood vessels beneath the skin

Cyanosis—blue appearance indicates poorly oxygenated hemoglobin, sign of hypoxia

Erythema—redness from increased blood flow to the skin due to emotional response (embarrassment) or inflammation

Jaundice—yellowing may indicate an increase in bilirubin in the blood, signifying decreased liver function

Pallor—paleness may be the result of an emotional response (fear or anger) or may indicate low blood pressure

Tan or Dark—increased melanin production may be the result of heredity or of increased exposure to UV light from the sun; this is a protective function to prevent UV damage to skin cells' DNA

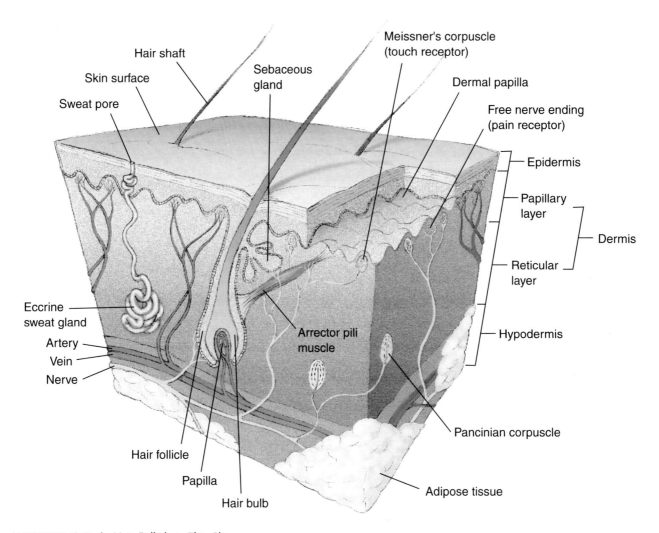

FIGURE 6.2 ◆ Hair Follicle in Thin Skin

◆ **Exercise 6.1** *Identify the Layers of Skin and Accessory Structures on a Skin Model*

Identify the following structures on a model of skin (Photo 356a). Review the functions of each structure as you identify it.

- Epidermis
- Dermis
- Hypodermis
- Hair follicle
- Hair bulb
- Hair (root and shaft)
- Sebaceous gland
- Arrector pili muscle
- Eccrine gland

◆ **Exercise 6.2** *Examine Microscopic Skin Structure*

A. Thin Skin Structure

Examine a thin skin slide and Photo 356b, and identify the following structures: *epidermis* (including each stratum, *keratinocytes,* and *melanocytes*), *dermis* (including *papillary* and *reticular layers*), and *hypodermis.* Sketch and label what you see.

B. **Thick Skin Structure**

Examine a thick skin slide (palmar skin) and Photo 357a, and identify the following structures: *epidermis* (including each stratum), *dermal layers* and the *hypodermis*. What additional stratum is present in the epidermis of thick skin?

C. **Comparison between Dark Skin (pigmented skin) and White Skin (non-pigmented skin)**

Examine the composite skin slides (pigmented compared to non-pigmented) and Photo 357b. Sketch and note the structural difference between them in the space below.

◆ Exercise 6.3 *Observe Hairs and Hair Follicles Under the Microscope*

Examine a slide of skin rich in hairs and hair follicles (*e.g.*, skin from scalp), and identify the following structures: *shaft, root, bulb, hair follicle, papilla* and *sebaceous gland*. Use Photos 358a–358d for reference. Sketch and label what you see in the space below.

Name _____

Lab 6: The Integumentary System

◆ **Practice**

1. What substance is manufactured in the skin (but is not a secretion) to play a role elsewhere in the body?

2. What sensory receptors are found in the skin?

3. A nurse tells a doctor that a patient is cyanotic. What is cyanosis? What does its presence imply?

4. How does sunlight cause pale skin to tan?

5. How does the skin help in regulating body temperature? (Describe two mechanisms.)

6. Match the skin colors below with the possible cause:

erythema	increased production of melanin
bronzing	low blood pressure
tanning	decreased production of hormones from the adrenal medulla
jaundice	inflammation
pallor	liver dysfunction

7. Label the diagram of a hair follicle:

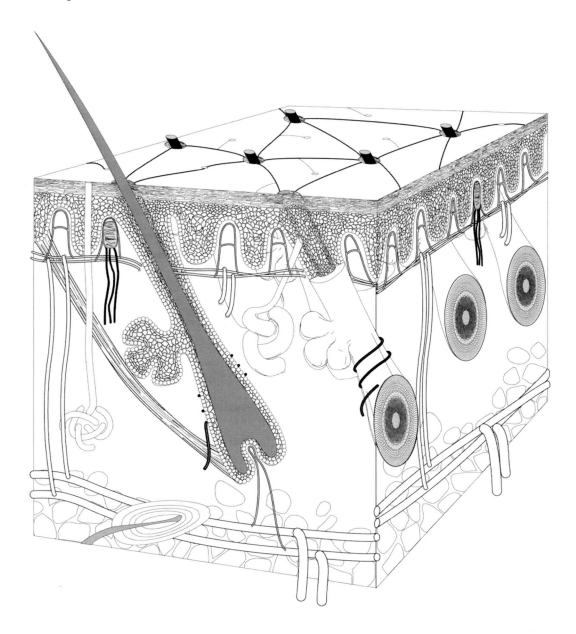

OSSEOUS TISSUE

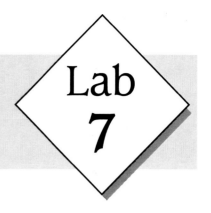

Lab
7

Objectives

1. Locate and identify the three types of cartilages on models, diagrams and microscope slides
2. Identify surface bone features and functions
3. Identify the microscopic structure of compact bones
4. Identify the anatomical features of a cow long bone

Materials

- Skeleton model
- Microscope slides
  - Hyaline cartilage
  - Elastic cartilage
  - Fibrocartilage
  - Cross section of compact bone
- Sagittally cut beef joint with long bone

INTRODUCTION

Our skeletons are made of osseous tissue and cartilage. Cartilage has a higher water content than bone, and allows firmness with resilience. These properties allow cartilage to function well in joint formation, protection of bone surfaces at joints (articulations), shock absorption and in maintaining shape with some flexibility. Cartilage, unlike bone, is avascular. Bone is more rigid than cartilage and allows for greater support and protection, but does not have the flexibility of cartilage. Bone is also the site of mineral storage, particularly for calcium and phosphate. The bone marrow is the site of blood cell formation. Bones also serve as attachments for muscles and are the means of locomotion for our bodies. In this exercise you will examine the properties and locations of the three types of skeletal cartilages and the two types of bone. You will also begin to classify the various markings on bones and become familiar with their functions.

CONCEPTS

I. Cartilage

Cartilage is found in regions where slight movement coupled with support is desirable. Some types are also found in joints to protect the ends of bones, to increase articular surface area, and to provide shock absorption. Cartilage is avascular and contains no nerve fibers.

Table 7.1
Skeletal Cartilages

◆◆◆

Cartilage Type	Locations	Functions	Properties
Hyaline	the most predominant cartilage in the skeleton; nasal cartilages, trachea and bronchi (respiratory passageways), costal cartilages (ribs), ends of long bones	support with some resiliency	spherical chondrocytes with fine collagen fibers that are not visible in the light microscope
Elastic	external ear, epiglottis	maintain shape with great flexibility	spherical chondrocytes with collagen and elastic fibers; elastic fibers are visible in the light microscope
Fibrocartilage	intervertebral discs, pubic symphysis, menisci of the knee and some other joints	provide great tensile strength and act as a shock absorber	small chondrocytes nestled between thick bundles of collagen fibers

◆ Exercise 7.1 *Observe and Draw Microscopic Views of Cartilage*

Examine microscope slides of hyaline cartilage, fibrocartilage and elastic cartilage. Using your colored pencils, draw a small section of each cartilage and compare your drawing with the photographs in the photo atlas.

A. Hyaline Cartilage (Photo 359a)
  • Draw your observations in the space below. Label the chondrocytes.

B. Elastic Cartilage (Photo 359b)
  • Draw your observations in the space below. Label the chondrocytes and elastic fibers.

C. Fibrocartilage (Photo 359c)

- Draw your observations in the space below. Label the chondrocytes and collagen fibers.

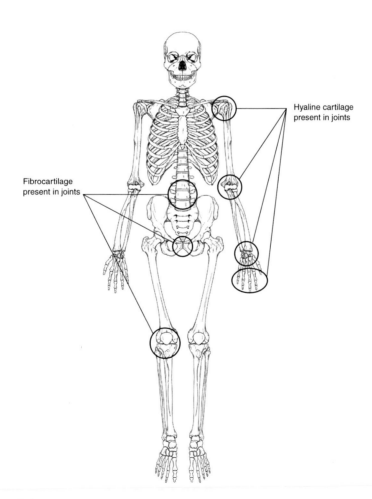

FIGURE 7.1 ◆ Anterior Skeleton

II. Bone

Bone is a denser and drier tissue that is more rigid than cartilage. Bone can be classified by the shape of the bone organ, or by the structure of the bone tissue.

A. Bone Shape

Table 7.2
Bone Shapes

◆◆◆

- **Long Bones**—length is longer than width; shaft of long bone is called the diaphysis; ends of the long bones are called the epiphyses (singular is epiphysis). See Figure 7.6.
  - Examples—limb bones, digits
- **Short Bones**—length and width are roughly the same
  - Examples—wrist bones, patella (knee cap)
- **Flat Bones**—thin, flat bones with some curvature
  - Examples—sternum (breast bone), ribs, skull bones
- **Irregular Bones**—Bones with complex processes or holes that can't be classified as any of the above shapes
  - Examples—vertebrae, coxal (hip) bones

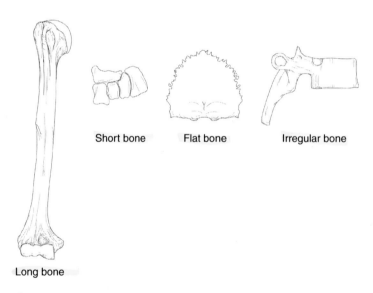

Short bone   Flat bone   Irregular bone

Long bone

FIGURE 7.2 ◆ Shapes of Bones

B. Bone Tissue Structure

Table 7.3
Bone Tissue Structure

◆◆◆

- **Compact Bone** — dense, homogenous tissue that is supplied by blood vessels and nerves through tunnel-like passageways
  - Locations — outside layer of all shapes of bones
- **Spongy Bone** — small bony struts and columns called trabeculae that have the appearance of a sponge, although they are very rigid; the space around the trabeculae is filled with red or yellow bone marrow
  - Locations — epiphyses of long bones, center of short bones and flat bones

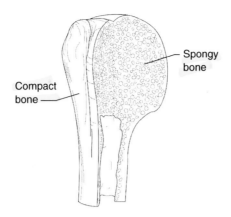

FIGURE 7.3 ◆ Compact and Spongy Bone in a Long Bone

C. Microscopic Anatomy of Compact Bone (Figures 7.4, 7.5)

The structural unit of compact bone is the **osteon.** The osteon is positioned vertically in bone and has a long vertical tunnel called the **central canal** running in the middle, that contains blood vessels and nerve fibers. The vessels and fibers reach the central canal through horizontal **perforating** canals extending from the surface of the bone. Surrounding the central canal are concentric non-living circles of bony matrix called **lamellae.** Tiny caverns between the lamellae, called **lacunae,** contain living **osteocytes** that contribute to the maintenance of the bony matrix. Blood supply to the osteocytes arrives through tiny canals extending from the central canal to the lacunae. These canals are called **canaliculi.** The strength of compact bone comes from the presence of collagen fibers and mineral salts called hydroxyapetites.

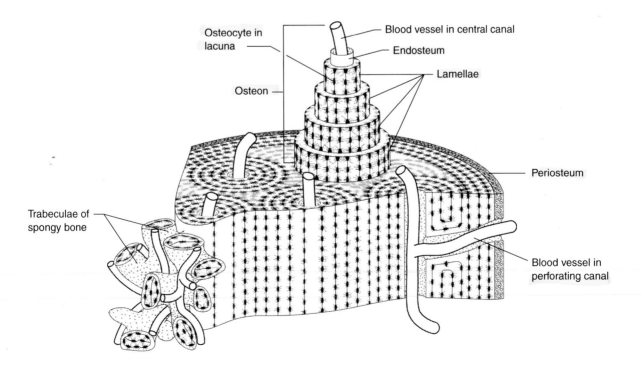

FIGURE 7.4 ◆ Structure of Compact Bone

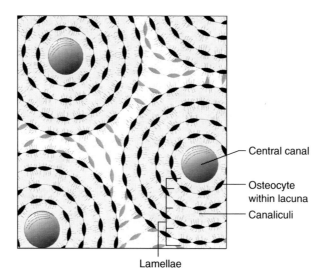

FIGURE 7.5 ◆ Microscopic View of Compact Bone

◆ **Exercise 7.2** *Observe and Draw a Microscopic Cross Section Through Compact Bone*

Examine slides of compact bone cross sections. Be able to identify the following components of an osteon on the slide, and explain the *functions* of these components: *lamellae, lacuna, osteocytes, canaliculi, perforating canal, central canal.* Use Photos 360a and b as a reference.

Sketch and label what you see in the space below:

III. Bone Features

Bone features, (also called bone markings), are bumps, grooves, holes and projections that serve various functions. Some bumps are sites for muscle attachments, while some projections form joints with other bones. Holes are often passageways for blood vessels and nerves. Table 7.4 lists bone features that you will find helpful to know as you study the skeleton in subsequent labs.

IV. Gross Anatomy of a Long Bone

A long bone (Figure 7.6) is composed of a long shaft, the **diaphysis,** between two **epiphyses.** The diaphysis is surrounded by a membrane called the **periosteum,** which is attached to the bone by fibrous connections called **perforating fibers.** Blood vessels and nerve fibers enter the periosteum through **nutrient foramina,** or holes within the membrane. These vessels and fibers then enter perforating canals in the compact bone of the diaphysis. In the center of the diaphysis is a long **medullary cavity** filled with yellow bone marrow. The epiphyses are spongy bone wrapped in compact bone and covered at the ends by hyaline cartilage. Red bone marrow fills the spaces in the spongy bone. The **epiphyseal line** is a bar of hyaline cartilage that bisects each epiphysis. This line is the remnant of the epiphyseal plate, or growth plate, that allows growth in length of the long bone until about puberty.

Table 7.4
Bone Features

Feature	Description/Function	Example
Tuberosity	Large rounded projection; site for muscle attachment	Tibial Tuberosity
Crest	Prominent ridge of bone; site for muscle attachment	Iliac crest in the coxal bone
Trochanter	Large, blunt, irregularly shaped process; site for muscle attachment	Greater and lesser trochanters of the femur
Tubercle	Small rounded projection; site for muscle attachment	Greater and lesser tubercle of the humerus
Epicondyle	Raised ridge above a condyle; site for muscle attachment	Epicondyles of the humerus
Spine	Sharp, peaked ridge; site for muscle attachment	Spine of the scapula
Head	Rounded prominence next to a narrow neck region; used in formation of joint	Head of the humerus
Condyle	Rounded projection involved in formation of a joint	Mandibular condyles
Foramen	Round or oval hole through bone; passageway for blood vessels and nerves	Foramen magnum of occipital bone
Fossa	Shallow depression; site of muscle attachment or articular surface	Subscapular fossa of the scapula
Meatus /Canal	Tunnel-like passageway; opening to another structure	External auditory meatus of the temporal bone

◆ **Exercise 7.3** *Examine a Cow Knee Joint and Long Bone*

Examine a cow knee joint (Photo 360c), which is an articulation between two long bones. Identify the following structures:

1. spongy bone with trabeculae
2. compact bone
3. diaphysis
4. epiphyses
5. periosteum with perforating fibers
6. medullary cavity with yellow marrow
7. articular cartilage (hyaline cartilage)
8. meniscus (fibrocartilage)

◆◆◆

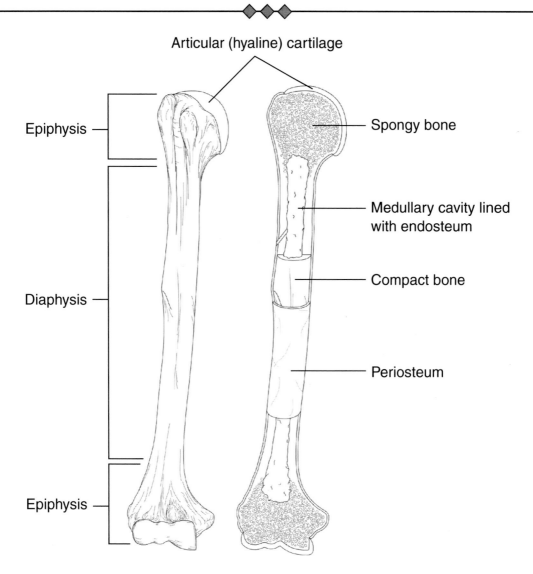

Articular (hyaline) cartilage

Epiphysis

Spongy bone

Medullary cavity lined with endosteum

Compact bone

Diaphysis

Periosteum

Epiphysis

FIGURE 7.6 ◆ Anatomy of a Long Bone

Name _____

Lab 7: Osseous Tissue

◆ Practice

1. Select from the following choices for each bone feature description.

   a. condyle
   b. crest
   c. epicondyle
   d. foramen
   e. fossa
   f. head
   g. meatus
   h. trochanter
   i. tubercle
   j. tuberosity
   k. spine

   __k.__ sharp, slender ridge is a site for muscle attachment

   __i__ small rounded projection that is a site for muscle attachment

   __b.__ prominent ridge of bone is a site for muscle attachment

   __j__ large rounded projection used as a site for muscle attachment

   __f.__ structure supported on neck

   __a.__ rounded projection used to form joints

   __g-__ canal-like structure

   __d.__ opening through a bone

   __e.__ shallow depression

   __h.__ large, irregularly shaped projection

   __c__ raised area above a condyle

2. Select from the following choices to fill in the blanks below.

   a. central canals
   b. osteocytes
   c. periosteum
   d. perforating canals
   e. canaliculi

   Blood supply and nutrients enter compact bone through nutrient foramina in the __c.__ Blood vessels travel vertically in compact bone through __a__ and horizontally through __d__. To reach the lacunae and their resident __b__, blood must move through vessels within the very tiny __c.__.

3. Choose the appropriate type of cartilage for each description below.

   a. elastic
   b. fibrocartilage
   c. hyaline

   _A_ supports the external ear

   _B_ between the vertebral bodies

   _C._ forms the walls of the trachea

   _C._ articular cartilages

   _B._ meniscus in a knee joint

   _C._ connects the ribs to the sternum

   _B_ most effective at resisting compression

   _A._ most springy and flexible

   _C._ most abundant

4. Label the diagram below:

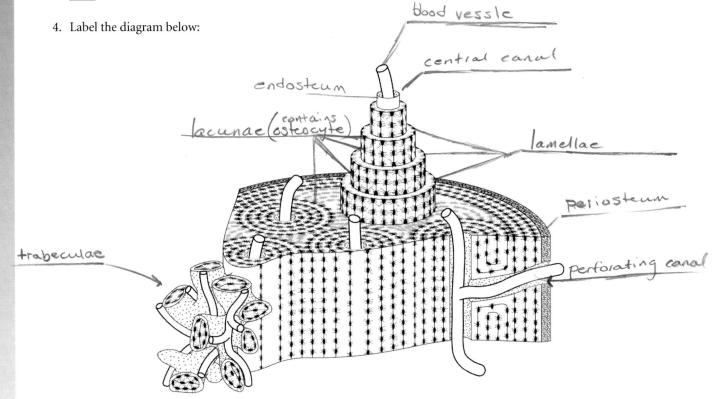

5. Label the diagram below:

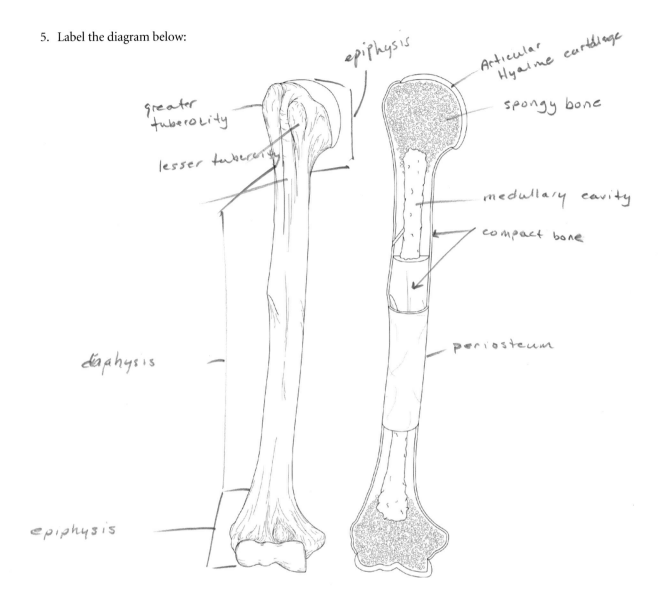

epiphysis

greater
tuberoLity

lesser tuberosity

Articular
Hyaline cartilage

spongy bone

medullary cavity

compact bone

periosteum

diaphysis

epiphysis

AXIAL SKELETON

Objectives

1. Identify the bones of the axial skeleton
2. Identify some important bone features on specific bones and their functional significance
3. Explain the significance of spinal curvatures and describe abnormal curvatures
4. Describe the structure and function of intervertebral discs

Materials

- Disarticulated bones of the axial skeleton
- Articulated skeleton
- Fetal skull model

INTRODUCTION

The bones of the skeleton are classified according to their position on the **axial** or the **appendicular** skeleton. *Axial* refers to the center (axis) of the body. Bones of the axial skeleton include the bones of the skull, the spinal column, ribs and sternum, all of which provide protection for softer, vital organs (Figure 8.1). The cranium of the skull surrounds and protects the brain. The vertebrae of the spinal column surround and protect the spinal cord. The thoracic cavity, bounded by the vertebral column, ribs and the sternum, protects the heart and lungs. In this lab you will examine both disarticulated bones and an articulated skeleton to become familiar with the bones of the axial skeleton.

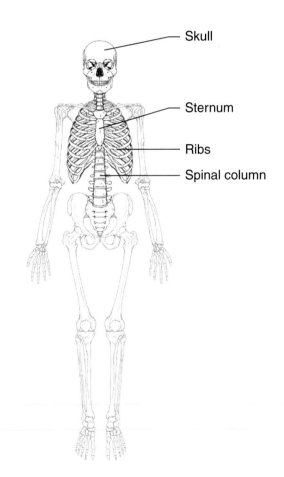

FIGURE 8.1 ◆ Axial Skeleton

CONCEPTS

I. Skull

The skull is a complex structure that consists of the **cranium,** which surrounds and contains the brain, and the **facial bones.**

The Infant Skull

Infant skulls have the same bones that are present in an adult skull; however, the proportions of the bones are different. One significant difference is that the fetal skull has membranous spaces between the cranial bones called **fontanels.** The anterior fontanel is the "soft spot" that you can feel in a baby's head. These spaces allow the cranial bones to slide over each other slightly to ease passage through the birth canal, and also allow for rapid brain growth for the first 18 months. The fontanels will undergo ossification through about the age of 20 months. The names of the fontanels are related to their location.

- **Anterior (frontal)** – at the junction between the frontal and both parietal bones
- **Mastoid (posterolateral)** – at the junction between the occipital, parietal and temporal bones
- **Sphenoid (anterolateral)** – at the junction between the frontal, parietal, sphenoid and temporal bones
- **Occipital (posterior)** – at the junction between the occipital and both parietal bones

Table 8.1
Bones of the Skull

◇◇◇

Cranium

Bone	# Present in Skull	Features
Frontal Bone	1	**Supraorbital foramen** is a passageway for blood vessels and nerves serving the forehead region
Parietal Bones	2	
Temporal Bones	2	**Zygomatic process** articulates with the zygomatic bone
		Mandibular fossa forms the temperomandibular joint (TMJ) with the mandible
		Mastoid process is an attachment site for the sternocleidomastoid muscle
		External auditory meatus is the ear canal that leads to the ear drum
Occipital Bone	1	**Foramen magnum** is a large passageway for the brainstem to exit the skull
		Occipital condyles articulate with the C_1 vertebra
Sphenoid Bone	1	**Optic canals** form a passageway for the optic nerves to carry vision information
		Foramen rotundum allows the maxillary branch of the 5th cranial nerve (Trigeminal nerve) to pass out of the skull; dentists inject anesthetic near this foramen to numb the upper jaw
		Sella turcica is named for its resemblance to a Turkish saddle; this region of the sphenoid bone houses the pituitary gland of the brain
Ethmoid Bone	1	**Cribriform plate** contains olfactory foramina through which olfactory nerves pass to the olfactory bulbs that rest on the cribriform plate

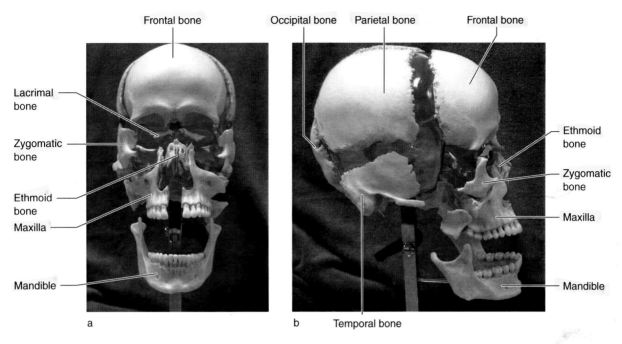

a b Temporal bone

FIGURE 8.2 ◆ Disarticulated Skull (Note: nasal bones and vomer not present)

Crista galli is a ridge of bone in the center of the cribriform plate that separates the olfactory bulbs
Middle nasal conchae form the lateral part of the nasal cavity; increase cavity surface area

Facial Bones

Bone	# Present in Skull	Features
Maxillae	2	**Palatine process** forms the anterior majority of the hard palate **Infraorbital foramen** allows passage of blood vessels and nerves that supply the upper jaw region
Mandible	1	**Mandibular condyle** forms the temperomandibular joint with the mandibular fossa of the temporal bone **Coronoid process** provides an attachment site for the temporalis muscle **Mandibular foramen** is a passageway for the mandibular division of the 5th cranial nerve (Trigeminal nerve); dentists inject anesthesia near here to numb the lower jaw **Body** forms the chin region **Mental foramen** is a passageway for blood vessels and nerves that supply the chin region
Zygomatic bone	2	**Temporal process** articulates with the temporal bone to form the zygomatic arch (your cheekbones)
Palatine bones	2	**Horizontal plate** forms the posterior part of the hard palate
Lacrimal bones	2	
Nasal bones	2	
Vomer	1	Forms part of the nasal septum
Inferior Nasal Conchae	2	Form the lateral part of the nasal cavity inferior to the middle nasal conchae

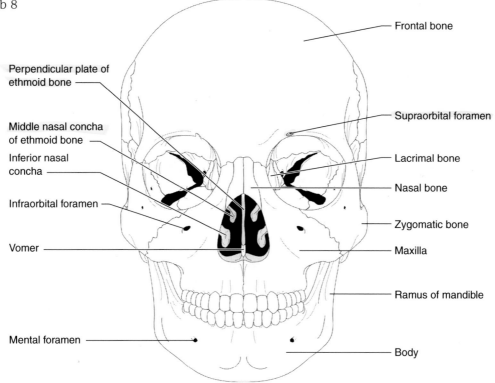

FIGURE 8.3 ◆ Facial Bones

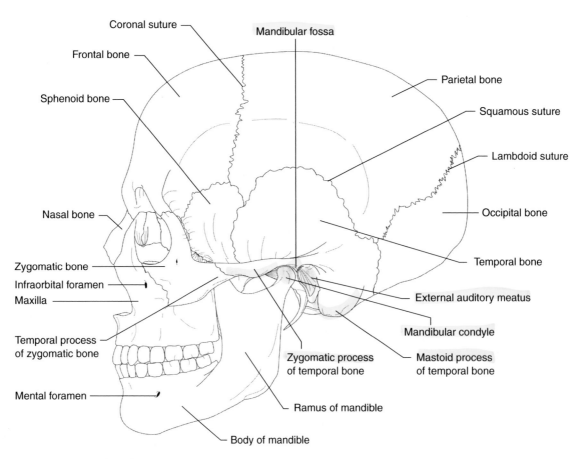

FIGURE 8.4 ◆ Lateral Skull

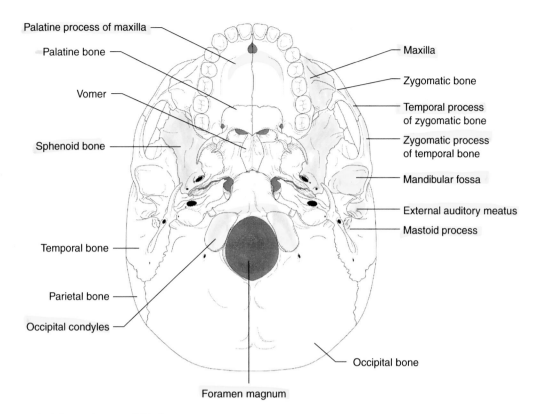

FIGURE 8.5 ◆ Inferior Aspect of Skull

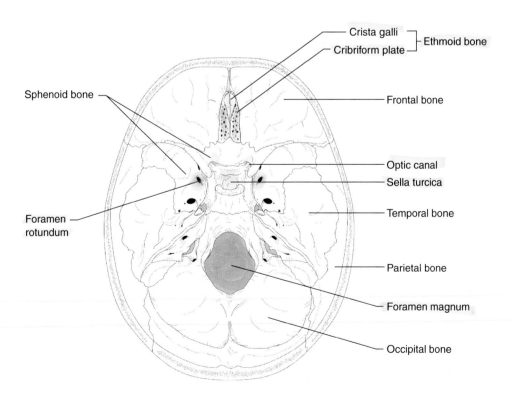

FIGURE 8.6 ◆ Floor of Cranial Vault

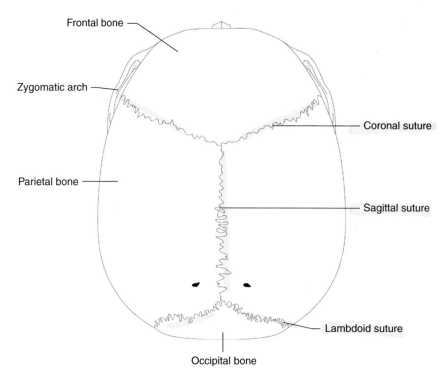

Frontal bone

Zygomatic arch

Coronal suture

Parietal bone

Sagittal suture

Lambdoid suture

Occipital bone

FIGURE 8.7 ◆ Sutures

Sutures

Sutures are fibrous joints between the bones of the cranium. The major sutures formed at the margins of the parietal bone are shown in Figures 8.4 and 8.7.

Hyoid Bone

The **hyoid bone** is located at the superior margin of the larynx (voicebox) and provides an attachment site for many tongue and neck muscles. It is unique in that it is the only bone in the body that does not articulate with any other bone.

II. Vertebral Column

The **vertebral column** is formed primarily from a stack of vertebrae separated by intervertebral discs. At the inferior end of the column is the sacrum and the small coccyx (tail bone).

Vertebrae

The vertebrae are divided into three regions. The **cervical** region consists of seven cervical vertebrae, the **thoracic** region has 12 vertebrae and the **lumbar** region contains five stocky vertebrae. The vertebrae are named with the first letter of the region in which they are located along with the number counted from the superior end. Thus, the cervical region has vertebrae C_1–C_7, thoracic region has T_1–T_{12}, and lumbar region contains L_1–L_5. Although vertebrae from each region have distinctive features, all vertebrae except C_1 and C_2 have a similar shape.

Articulated vertebral bodies are separated by a fibrocartilaginous **intervertebral disc.** The discs allow some flexion of the spine and provide a shock absorbing effect. The **vertebral foramina** of articulated vertebrae are aligned to form the **vertebral canal,** in which is located the spinal cord. Spinal nerves exit the spinal cord through the **intervertebral foramina,** that are more easily seen in articulated vertebrae.

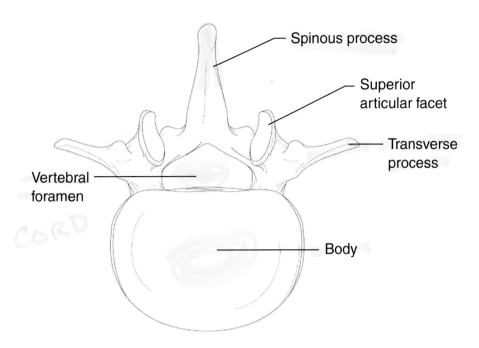

FIGURE 8.8 ◆ Typical Vertebra

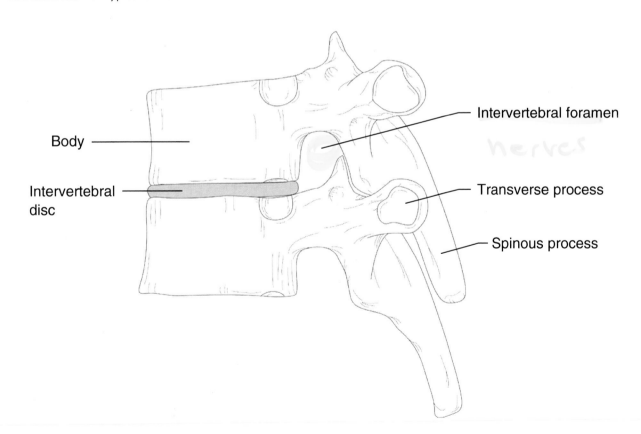

FIGURE 8.9 ◆ Articulated Vertebrae

Table 8.2
Unique Features of Specific Vertebrae

◆◆◆

- **Cervical Vertebrae**
  - **C1**—also named **Atlas** (from the Greek god who carried the sky on his shoulders); no body; no spinous process; superior articular surfaces form joints with the occipital condyles; inferior articular surfaces articulate with C2 and allow extension and flexion of the head to create a "yes" movement
  - **C2**—also named **Axis;** body projects superiorly to form the **dens,** or **odontoid process,** that projects superiorly through the atlas; the atlas rotates around the dens to allow a "no" movement of the head
  - **C7**—larger spinous process is easily palpated through the skin
  - **All cervical vertebrae**—transverse processes contain a transverse foramen through which pass the vertebral arteries to supply the brain
- **Thoracic Vertebrae**—two demifacets on the sides articulate with the heads of the ribs; long spinous processes point inferiorly
- **Lumbar Vertebrae**—larger vertebral bodies; short, flat spinous processes

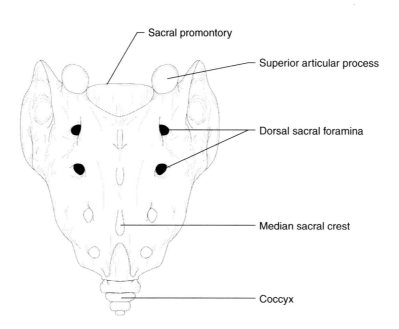

FIGURE 8.10 ◆ Dorsal Sacrum

Sacrum

The **sacrum** is formed from five fused vertebrae. The superior articular processes articulate with L5. Laterally, the **auricular surfaces** articulate with the coxal (hip) bones to form the sacroiliac joint. The **sacral promontory** is a ridge of bone at the anterior aspect of the superior articular surface. Spinal nerves exit through the **dorsal sacral foramina** on the dorsal surface of the sacrum. The **ventral sacral foramina** are passageways for blood vessels and nerves.

Coccyx

The **coccyx** is the triangular shaped tailbone and serves no useful purpose.

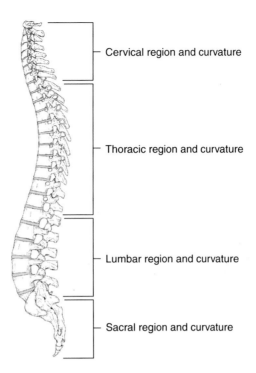

- Cervical region and curvature
- Thoracic region and curvature
- Lumbar region and curvature
- Sacral region and curvature

FIGURE 8.11 ◆ Curvatures and Regions of the Vertebral Column

Table 8.3
Abnormal Curvatures of the Spine

- Scoliosis—lateral curvature, usually in the thoracic region; can compromise breathing functions
- Kyphosis—also known as "hunchback"; dorsal curvature in the thoracic region
- Lordosis—also known as "swayback"; exaggerated lumbar curvature; may exist temporarily in pregnant women

Vertebral Curvatures

From a lateral view the adult vertebral column exhibits four curvatures. The resultant **S** shape allows more flexibility than a rigid, straight spine. The cervical and lumbar curvatures bend anteriorly, while the thoracic and sacral curvatures bend posteriorly.

Intervertebral Discs

The Frisbee shaped intervertebral discs are composed of an outer rim of fibrocartilaginous **annulus fibrosus** and a gelatinous inner **nucleus pulposus.** The ability to bulge laterally gives the intervertebral discs shock absorbing properties, as well as protective support between vertebrae. If a disc herniates (protrudes abnormally, usually to one side), the adjacent vertebrae may lose their alignment and may pinch a spinal nerve as it exits through the intervertebral foramina. Such an impingement can cause pain and loss of function.

III. Bony Thorax (Thoracic Cage)

The sternum (breastbone), ribs and costal cartilages together make up the bony thorax, which is rigid enough to protect the heart and lungs, but flexible enough to allow expansion for breathing.

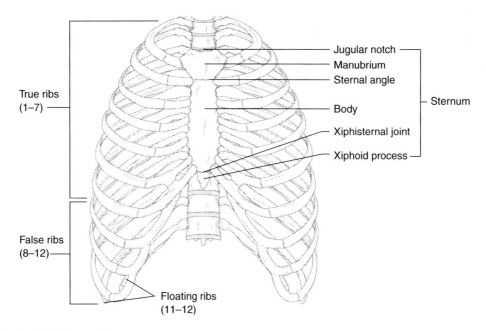

FIGURE 8.12 ◆ The Thoracic Cage

Sternum

The **sternum** is divided into three regions. The **manubrium** is the roughly heart-shaped superior portion. The **jugular notch** is the small indentation in the center superior margin of the manubrium. The **body** of the sternum is the long, flat portion that we typically think of as the breastbone. It connects with the manubrium at the **sternal angle.** Extending from the inferior aspect of the body is the small **xiphoid process,** which articulates with the body at the **xiphisternal joint.**

Ribs

There are 12 pairs of **ribs** extending from the thoracic region of the vertebral column. The first seven ribs have a direct attachment to the sternum via hyaline cartilage and are called **true ribs.** The remaining 5 ribs do not have a direct connection to the sternum and are known as the **false ribs.** Hyaline cartilages that connect the ribs to the sternum, either directly or indirectly, are called the **costal** cartilages. Ribs 11 and 12 have no attachment to the sternum and are also called **floating ribs.**

◆ **Exercise 8.1**   *Identify Bones and Bone Features*

Using disarticulated bones and an articulated skeleton, identify all of the bones and bone features of the axial skeleton that are listed in **bold** type in this lab.

Name _____

Lab 8: Axial Skeleton

◆ **Practice**

1. The diagrams A, B, and C below show different views of the skull. Color the circles next to the name of each bone and then color the corresponding bone with the same color. Label the sutures and bone features designated with leader lines.

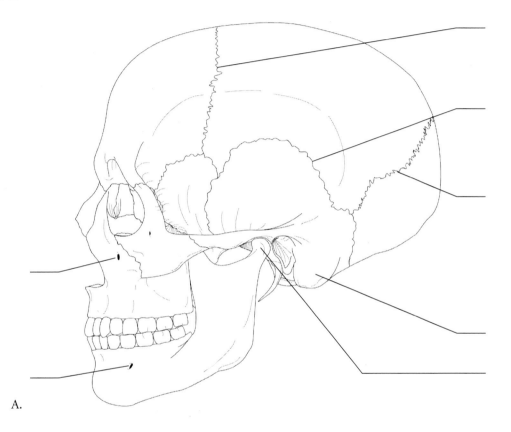

A.

○ Frontal   ○ Nasal

○ Parietal   ○ Maxilla

○ Occipital   ○ Mandible

○ Temporal   ○ Vomer

○ Zygomatic   ○ Lacrimal

○ Sphenoid   ○ Palatine

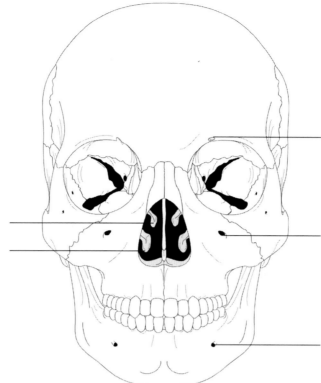

B.

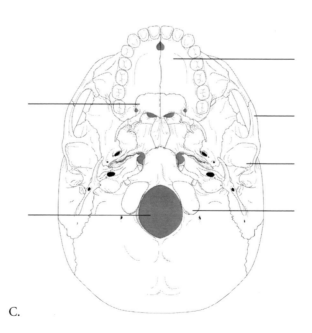

C.

2. Which bones form the hard palate? What do you think are the functions of the hard palate? (Hint: Think about eating and speaking.)

3. Which bones form the nasal septum?

4. What is a fontanel? What are the functions of fontanels in the fetal and newborn skull?

5. Choose from these options to correctly identify the vertebra or region of the vertebral column listed below.

   a. atlas
   b. axis
   c. cervical vertebra
   d. coccyx
   e. lumbar vertebra
   f. sacrum
   g. thoracic vertebra

   _____ vertebral type containing foramina in the transverse processes

   _____ dens provides a pivot for rotation of the 1st cervical vertebra

   _____ transverse processes faceted for articulation with ribs; spinous process pointing sharply downward

   _____ articulates with the hip bone laterally

   _____ massive vertebral body and blunt, flat spinous process

   _____ "tail bone"

   _____ articulates with the occipital condyles to support the head and allow a "yes" movement

   _____ seven vertebrae with an anterior curvature

   _____ twelve vertebrae with a posterior curvature

6. Describe how a spinal nerve exits from the vertebral column.

7. What are the functions of the intervertebral discs? What problems might result from a herniated disc?

8. What is the function of the spinal curvatures?

9. Describe the following abnormal spinal curvatures:

   a. scoliosis

   b. kyphosis

   c. lordosis

10. Which organs are protected by the thoracic cage?

11. Write the numbers of the ribs that match the descriptions in the blanks below.

   _____ false ribs

   _____ true ribs

   _____ floating ribs

   _____ direct attachment to the sternum

   _____ indirect attachment to the sternum

   _____ no attachment to the sternum

   _____ connected to costal cartilages

   _____ connected to thoracic vertebra posteriorly

12. In the table below write the name of the bone on which you find the indicated bone feature. Include the function of the particular bone feature.

Bone Feature	Bone	Function of Bone Feature
Infraorbital foramen		
Mastoid process		
Mandibular condyle		
Foramen rotundum		
External auditory meatus		
Mental foramen		
Vertebral foramen		
Auricular surface		
Dens		

APPENDICULAR SKELETON

Lab
9

Objectives

1. Identify the bones of the appendicular skeleton
2. Identify some important bone features on specific bones and their functional significance

Materials

- Disarticulated bones of the appendicular skeleton
- Articulated skeleton

◆◆◆

INTRODUCTION

In Lab 8 we examined the bones of the axial skeleton. The **appendicular** skeleton consists of the appendages and their attachments to the axial skeleton. The upper extremity (arm) is attached to the axial skeleton by the **pectoral girdle,** while the lower extremity (leg) is connected via the **pelvic girdle.** The appendicular skeleton enables us to move, and supports us as we stand. In this lab you will become familiar with the appendicular skeleton by examining disarticulated bones and an articulated skeleton.

CONCEPTS

I. Pectoral Girdle

The pectoral girdle is connected to the axial skeleton mainly by muscle attachments to the ribs, sternum and vertebral column. The pectoral girdle consists of the **clavicle,** commonly called the collarbone, and by the **scapula,** or shoulder blade. The clavicle articulates at its medial end to the sternum and laterally with the scapula. The scapula articulates with the humerus (arm bone) in addition to the clavicle. Most of the scapula rests on the posterior surface of the thorax. The unique structure of the pectoral girdle allows tremendous freedom of movement of the upper extremity.

Table 9.1
Bones of the Pectoral Girdle

◆◆◆

- **Clavicle—sternal end** articulates medially with the sternum; **acromial end** articulates laterally with the scapula
- **Scapula** (shoulder blade)
  - **acromion process** articulates with the clavicle to form acromioclavicle joint
  - **coracoid process** is an attachment site for the biceps brachii, brachialis, and pectoralis minor muscles
  - **glenoid fossa** (glenoid cavity) articulates with the head of humerus to form the shoulder (glenohumeral) joint
  - **spine** provides an attachment site for muscles of the shoulder
  - **supraspinous fossa** is an attachment site for the supraspinatus muscle
  - **infraspinous fossa** is an attachment site for the infraspinatus muscle
  - **subscapular fossa** is an attachment site for the subscapularis muscle

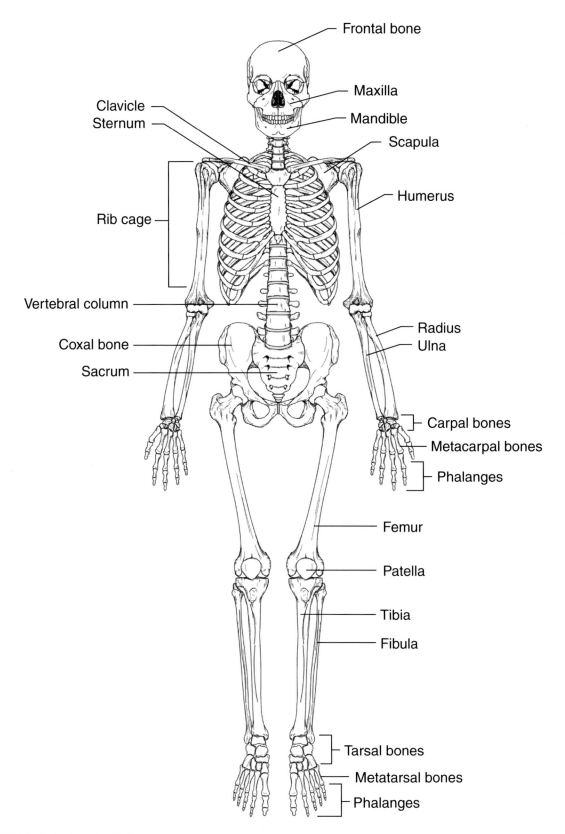

FIGURE 9.1 ◆ Anterior Skeleton

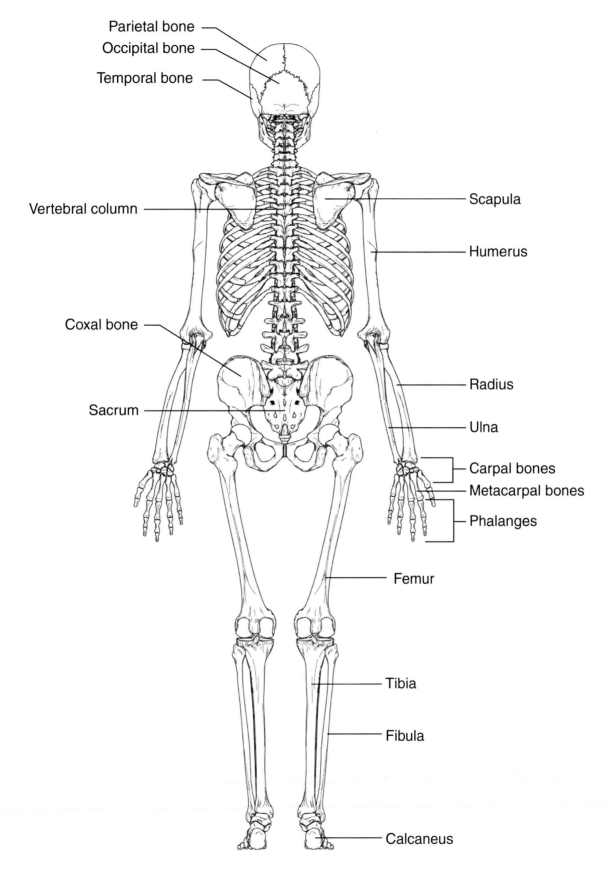

Parietal bone

Occipital bone

Temporal bone

Vertebral column

Scapula

Humerus

Coxal bone

Sacrum

Radius

Ulna

Carpal bones

Metacarpal bones

Phalanges

Femur

Tibia

Fibula

Calcaneus

FIGURE 9.2 ◆ Posterior Skeleton

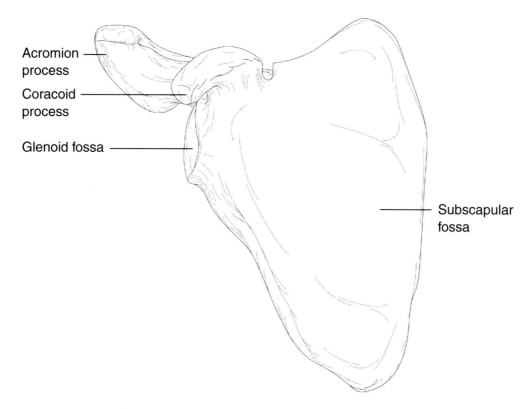

FIGURE 9.3 ◆ Anterior Scapula

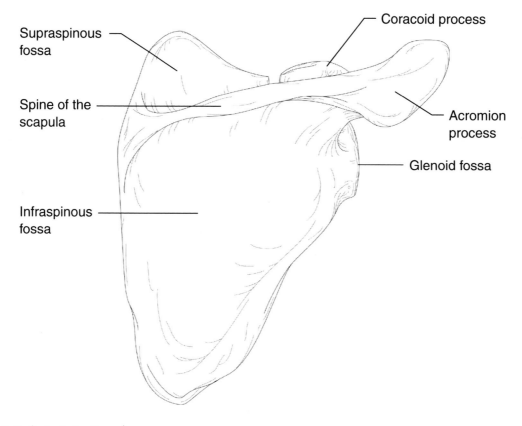

FIGURE 9.4 ◆ Posterior Scapula

II. **Upper Limb**

Table 9.2
Bones of the Upper Extremity

Humerus
- **head** articulates with the glenoid fossa of the scapula to form shoulder (glenohumeral) joint
- **greater and lesser tubercles** are sites of muscle attachment; the greater tubercle sits proximal and lateral to the lesser tubercle
- **intertubercular groove** is a passageway for the tendon of biceps brachii
- **anatomical neck** is a narrow region that supports the head
- **surgical neck** is the most frequently fractured part of the humerus
- **deltoid tuberosity** is an attachment site for the deltoid muscle
- **trochlea** articulates with the ulna to form part of the elbow joint
- **capitulum** provides a rounded surface for the head of the radius to slide over
- **olecranon fossa** fits the olecranon process of the ulna
- **coronoid fossa** fits the coronoid process of the ulna

Radius
- **head** is the proximal surface, which articulates with the capitulum of the humerus to form part of the elbow joint; the medial surface articulates with the radial notch of ulna to form the proximal radioulnar joint
- **radial tuberosity** is an attachment site for the biceps brachii
- **ulnar notch** articulates with the ulnar head to form the distal radioulnar joint
- **styloid process** provides lateral stability for the wrist joint; articulates with scaphoid bone of the wrist

Ulna
- **trochlear notch** articulates with the trochlea of the humerus to form a hinge joint at the elbow
- **olecranon process** stabilizes elbow joint posteriorly
- **coronoid process** stabilizes elbow joint anteriorly
- **radial notch** articulates with the radius head to form the proximal radioulnar joint
- **head** is the distal end, which articulates with the ulnar notch of radius to form the distal radioulnar joint
- **styloid process** provides medial stability for the wrist joint; articulates with the lunate bone of the wrist

Hand
- **Carpal bones**—bones of the carpus (wrist)
  - proximal row (from lateral to medial)
        scaphoid, lunate, triquetral, pisiform
  - distal row (from lateral to medial)
        trapezium, trapezoid, capitate, hamate
- **Metacarpals**—five per hand, from lateral to medial
      metacarpal 1, metacarpal 2, metacarpal 3, metacarpal 4, metacarpal 5
- **Phalanges**—three per digit (except thumb), from lateral to medial
      proximal phalanx 1, distal phalanx 1 (there is no middle phalanx 1),
      proximal phalanx 2, middle phalanx 2, distal phalanx 2,
      proximal phalanx 3, middle phalanx 3, distal phalanx 3,
      proximal phalanx 4, middle phalanx 4, distal phalanx 4,
      proximal phalanx 5, middle phalanx 5, distal phalanx 5

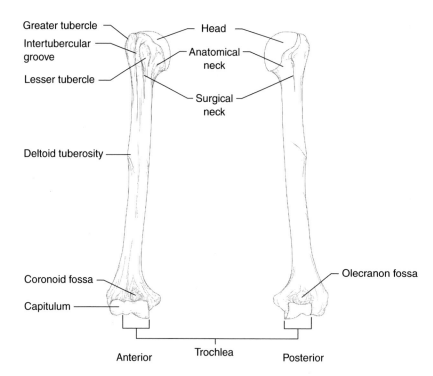

FIGURE 9.5 ◆ Anterior and Posterior Views of the Humerus

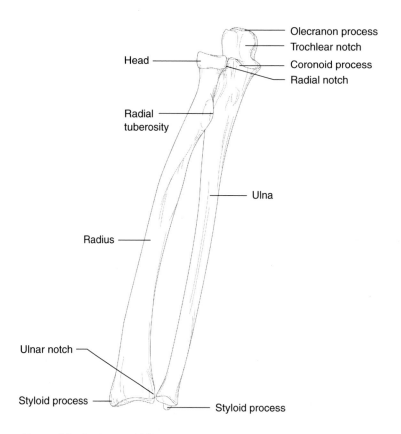

FIGURE 9.6 ◆ Anterior View of the Radius and Ulna

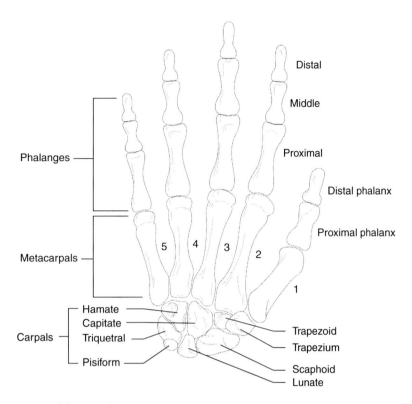

FIGURE 9.7 ◆ Anterior View of the Hand

III. Pelvic Girdle

The pelvic girdle transfers the weight of the upper body to the lower extremities, and supports the pelvic organs. The paired **coxal bones** form the hips, and are secured to the sacrum posteriorly. Together the coxal bones and the sacrum form the **pelvis**. The lower extremities experience less freedom of movement and more stability than the upper extremities.

Pelvic Cavity

The **pelvic inlet** is defined by the inner circular ridge at the level of the sacral promontory. The **pelvic outlet** is best seen on the inferior view of the pelvis, and includes the ischial tuberosities and coccyx posteriorly, and the pubis regions anteriorly. Male and female pelvises structurally reflect differences in childbearing capacity and in build (Figure 9.10). The female coxae are flared laterally to widen the pelvic inlet, or birth canal. The sacrum and coccyx are positioned more posteriorly to provide a wider pelvic outlet. The male coxae, on the other hand, are positioned more vertically and the pelvic inlet is narrower. The sacrum and coccyx curve anteriorly so that the pelvic outlet is also more narrow than in the female. The male pelvis is better adapted to support a heavier build, while the female pelvis is shaped to permit passage of a baby.

Table 9.3
Bones of the Pelvic Girdle

◆◆◆

Coxal Bone—formed from three fused regions:

1. **Ilium** (superior portion of the coxal bone)
   - **iliac crest** is the superior most edge upon which you rest your hands on the hips
   - **greater sciatic notch** fits around the sciatic nerve as it passes beside the coxae
   - **acetabulum** is the articular socket that receives the head of the femur; all three regions of the coxal bones contribute to the acetabulum
   - **auricular surface** articulates with the auricular surface of the sacrum to form the sacroiliac joint; this joint becomes loose during labor
2. **Ischium** (inferior and posterior part of the coxal bone)
   - **lesser sciatic notch** is a passageway for blood vessels and nerves that serve the perineum
   - **ischial tuberosity** is a thick, roughened protrusion that bears weight when sitting
3. **Pubis** (anterior portion of the coxal bone)
   - **obturator foramen** is closed by the obturator membrane, but allows some blood vessels and nerves to pass through
   - **pubic symphysis** is a fibrocartilaginous articulation between the two coxal bones; this joint becomes loose during labor

Sacrum—See page 88

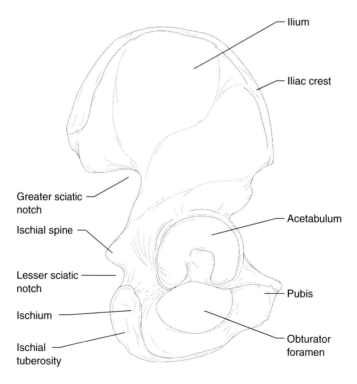

FIGURE 9.8 ◆ Lateral View of Coxal Bone

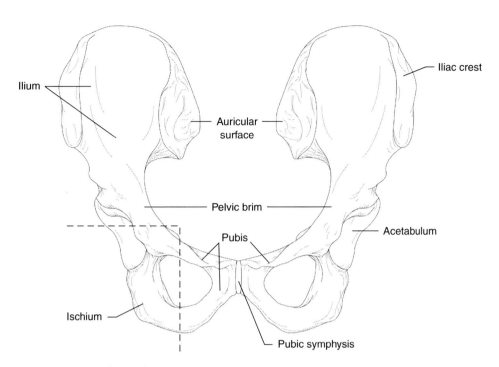

FIGURE 9.9 ◆ Anterior View of Coxal Bones

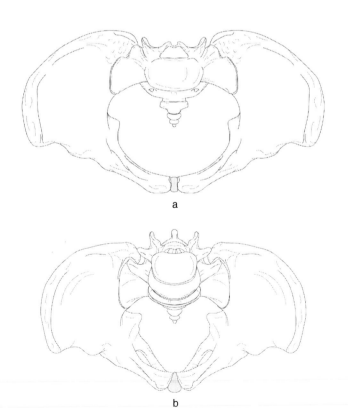

FIGURE 9.10 ◆ (a) Female and (b) Male Pelvis

IV. Lower limb

Table 9.4
Bones of the Lower Extremity

◆◆◆

Femur
- **head** articulates with the acetabulum of the coxal bone
- **neck** is a narrow region positioned obliquely to the head; the most frequently fractured part of the femur
- **greater and lesser trochanters** are attachment sites for thigh and buttock muscles; the greater trochanter is positioned proximally and laterally to the lesser trochanter
- **lateral and medial condyles** articulate with the tibial head to form the hinge joint of the knee
- **intercondylar notch** is a gap that houses the anterior and posterior cruciate ligaments

Patella (knee cap) protects the knee joint and increases the leverage of the patellar tendon of the quadriceps as it extends the knee joint

Tibia
- **medial and lateral condyles** articulate with the medial and lateral condyles of the femur to form the knee joint
- **tibial tuberosity** serves as an attachment site for the patellar ligament
- **medial malleolus** provides medial stability to ankle joint; articulates with the talus bone of the ankle

Fibula—is not a weight bearing bone, but provides attachment site for muscles and stabilizes the ankle joint
- **head** articulates with the lateral condyle of the tibia to form the proximal tibiofibular joint
- **lateral malleolus** provides lateral stability to the ankle joint; articulates with the talus bone of the ankle

Foot
> **Tarsal bones**—bones of the tarsus (ankle):
>> **talus, calcaneus, navicular, medial cuneiform, intermediate cuneiform, lateral cuneiform, cuboid**
>
> **Metatarsals**—five per foot, from medial to lateral:
>> **metatarsal 1, metatarsal 2, metatarsal 3, metatarsal 4, metatarsal 5**
>
> **Phalanges**—three per toe (except hallux), from medial to lateral:
>> **proximal phalanx 1, distal phalanx 1** (there is no middle phalanx 1),
>> **proximal phalanx 2, middle phalanx 2, distal phalanx 2,**
>> **proximal phalanx 3, middle phalanx 3, distal phalanx 3,**
>> **proximal phalanx 4, middle phalanx 4, distal phalanx 4,**
>> **proximal phalanx 5, middle phalanx 5, distal phalanx 5**

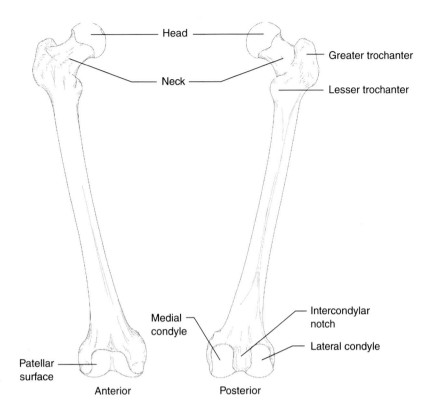

FIGURE 9.11 ◆ Anterior and Posterior View of the Femur

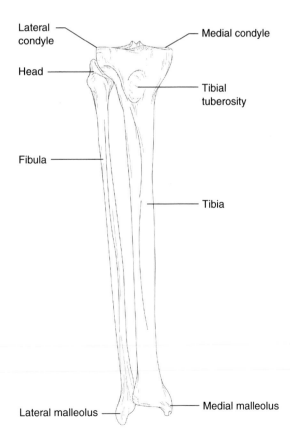

FIGURE 9.12 ◆ Anterior View of the Tibia and Fibula

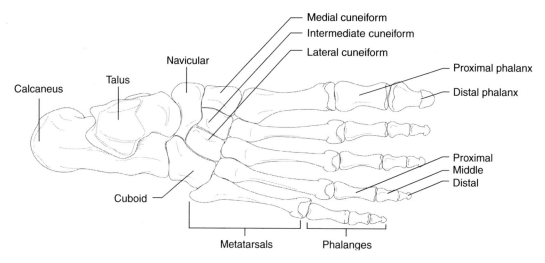

FIGURE 9.13 ◆ Superior View of the Foot

◆ Exercise 9.1 *Identify Bones and Bone Markings*

Using disarticulated bones and an articulated skeleton, identify all of the bones and bone features of the appendicular skeleton that are listed in **bold** type in this lab.

◆◆◆

Name _____

Lab 9: Appendicular Skeleton

 Practice

1. Label the structures indicated by leader lines on the diagrams.

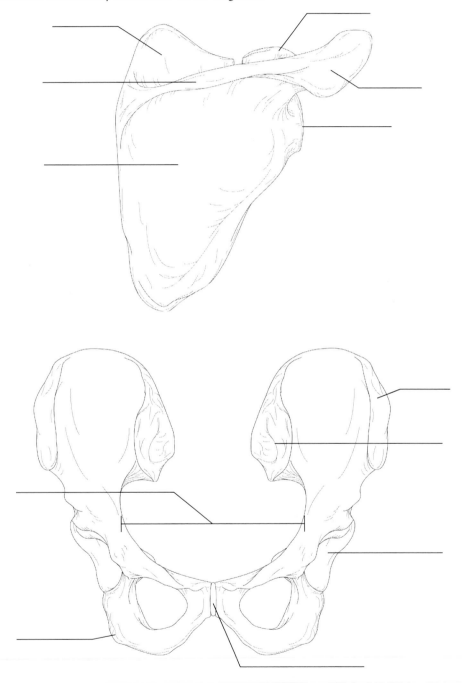

2. In the table below write the name of the bone on which you find the indicated bone feature. Include the function of the particular bone feature.

Bone Feature	Bone	Function of Bone Feature
Olecranon process		
Tibial tuberosity		
Intercondylar notch		
Sciatic notch		
Acromion process		
Intertubercular groove		
Greater trochanter		
Lateral malleolus		
Supraspinous fossa		
Pubic symphysis		

3. In the table below are listed several bone features involved in articulations with other bones. Fill in the name of the bone on which you find the listed feature, the articulating bone and the articulating bone feature if known.

Bone Feature	Bone	Articulating Bone	Articulating Bone Feature
Capitulum	Humerus	Radius	Head of radius
Styloid process of radius			None
Glenoid cavity			
Auricular surface of coxal bone	Coxal bone		
Condyles of femur			
Medial Malleolus			None
Trochlear notch			
Coronoid process of ulna			
Acetabulum			

4. List the differences between the male and female pelvis.

5. How do the functions of the pectoral girdle compare with those of the pelvic girdle?

SYNOVIAL JOINTS

Lab
10

Objectives

1. Describe the characteristics of synovial joints
2. Explain the movements capable of synovial joints
3. Identify the structure of the following synovial joints: shoulder, elbow, knee
4. Identify specific structures in a sagittal section of a cow knee joint

Materials

- Joint models
  - Shoulder, Rotator Cuff
  - Elbow
  - Knee
- Sagittal section through beef knee joint

INTRODUCTION

Joints are articulations, or connections, between two bones. Although there are several classifications of joints, those with which we are most familiar are the **synovial** joints. These freely moveable joints have characteristic structures that may be assembled in different ways in specific joints. In this lab you will review the structure of four major synovial joints and examine a cow knee joint.

CONCEPTS

I. Synovial Joint Movements

The language of anatomy includes descriptive words for movements. Table 10.1 describes the more common movements of which synovial joints are capable.

Table 10.1
Synovial Joint Movements

Flexion and **Extension**	Decreasing or increasing the joint angle in a sagittal plane
Dorsiflexion and **Plantarflexion**	Pointing the toes superiorly toward the head or inferiorly away from the head
Abduction and **Adduction**	Moving an appendage away from or toward the midline in a frontal plane
Circumduction	Moving the distal aspect of an appendage in a circle so that the entire limb describes a cone shape
Rotation	Moving a bone around its own axis
Supination and **Pronation**	Moving the palms anteriorly or posteriorly
Inversion and **Eversion**	Turning the sole of the foot medially or laterally
Protraction and **Retraction**	Projecting a bone forward or backward in a horizontal plane
Elevation and **Depression**	Projecting a bone superiorly or inferiorly in a vertical plane

II. Characteristics of Synovial Joints

Table 10.2
Characteristics of Synovial Joints

◆◆◆

- **Joint Cavity** All synovial joints are surrounded by a protective, fluid-filled cavity.
- **Articular Capsule** The articular capsule is a double layered membrane that forms the joint cavity.
  - **Fibrous Capsule** is the outer layer of the articular capsule membrane. This is a tough layer of dense irregular connective tissue that is continuous with the periosteum.
  - **Synovial Membrane** is the inner layer of the articular capsule that is composed of areolar connective tissue. This layer covers all internal surfaces of the joint cavity except for the articular surfaces of the bones. The membrane secretes synovial fluid into the capsule.
- **Articular Cartilage** The articulating surfaces of bones are covered with a protective layer of hyaline cartilage, which also absorbs some compression.
- **Synovial Fluid** The synovial membrane secretes synovial fluid into the joint cavity to lubricate articulating surfaces. This viscous fluid seeps into the articulating cartilage when the joint is not under pressure, then oozes back out to the surface when under pressure. This movement of the fluid in and out of the cartilage is called **weeping lubrication.**
- **Reinforcing Ligaments** Synovial joints are stabilized by the presence of ligaments that connect the articulating bones. The ligaments prevent the bones from separating.
- Structures present in some synovial joints
  - **Menisci** are pads of fibrocartilage that deepen the articular surface of some joints, and provide a shock absorbing surface.
  - **Bursae** are fluid filled sacs that reduce the friction of soft tissues that would otherwise come into contact with the joint as it moves.
  - **Tendon Sheaths** are similar to bursae, but are elongated and wrap around tendons to protect them.

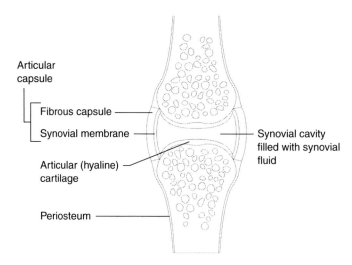

FIGURE 10.1 ◆ Synovial Joint

III. Specific Joints

Remember that, although synovial joints all have specific characteristics, each has a unique structure specially suited to its location and function. As you study the joints in this section, be aware of the features that increase joint stability and utility: the number and location of ligaments, the muscle tone within the tendons that cross the joint, and the depth and surface area of the articular surface.

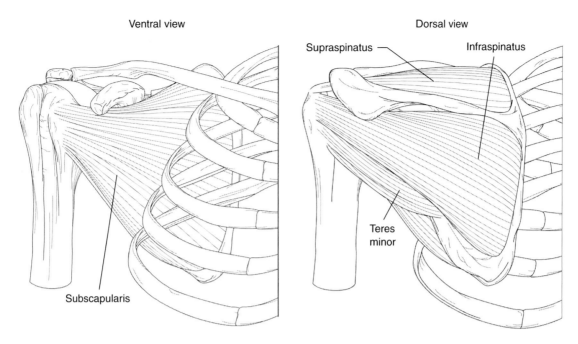

Ventral view   Dorsal view

Supraspinatus   Infraspinatus

Teres
minor

Subscapularis

FIGURE 10.2 ◆ Muscles of the Rotator Cuff

Shoulder

The shoulder is known as the **glenohumeral** joint. It is an articulation between the head of the humerus and the glenoid cavity of the scapula. This joint is classified as a **ball in socket** joint, even though the "socket" (glenoid cavity) is very shallow. The lack of depth and surface area of the articular surface cause this joint to tend to be unstable. That is, it may become dislocated more easily than some other joints. To improve the articular surface, a fibrocartilaginous ring called the **glenoid labrum** rims the glenoid cavity.

Some of the stabilizing ligaments present at the shoulder joint include the **coracohumeral ligament** and the **glenohumeral ligaments.** Both are part of the joint cavity. The coracohumeral ligament originates at the coracoid process of the scapula and inserts on the anterior head of the humerus. The glenohumeral ligaments, which may actually be absent in some individuals, weakly reinforce the anterior aspect of the capsule as they extend from the rim of the glenoid cavity to the head of the humerus.

The primary stabilizing feature of the shoulder joint is a group of four muscle tendons that form the **rotator cuff. Subscapularis** originates in the subscapular fossa and its tendon wraps medially to the head of the humerus to insert in the lateral aspect. **Supraspinatus** originates in the supraspinous fossa and **infraspinatus** originates in the infraspinous fossa. Tendons from both muscles wrap laterally around the head of the humerus to insert in the lateral aspect. **Teres minor** originates at the inferomedial border of the scapula and its tendon wraps laterally around the head of the humerus to insert in the lateral aspect.

Elbow

The elbow is a relatively stable hinge connection between the humerus and the ulna that permits flexion and extension of the arm. The primary stabilizing feature of this joint is the grasp of the ulnar trochlear notch around the humeral trochlea. The head of the radius glides over the surface of the capitulum of the humerus, but does not participate in, nor provide any stability to the elbow joint. Reinforcing ligaments that contribute to joint stability include the annular ligament and the ulnar and radial ligaments. The **annular ligament** forms a "sleeve" around the head of the radius, allowing it to rotate as the forearm supinates and pronates. The **ulnar (medial) collateral ligament** and **radial (lateral) collateral ligament** prevent lateral dislocation. Tendons of the biceps brachii and brachialis muscles cross the elbow anteriorly, while the triceps brachii muscle tendon supports the posterior aspect.

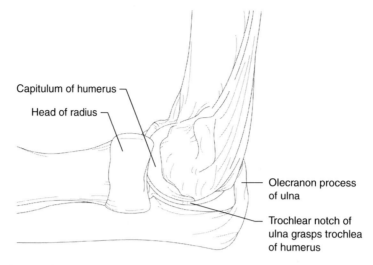

FIGURE 10.3 ◆ Lateral View of the Elbow Joint

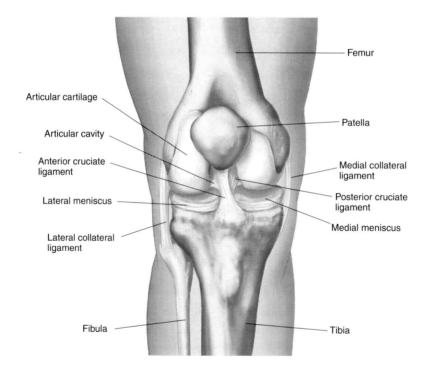

FIGURE 10.4 ◆ Anterior View of the Knee Joint

Knee

The most complex synovial joint in the body is the knee joint. Within one joint cavity are three articulations. The patella glides over the femur at the **femeropatellar joint.** Two femoral condyles connect with the tibial condyles and together are called the **tibiofemoral joint.** Together these articulations permit extension, flexion, and some rotation during flexion. Between the femoral and tibial condyles are two C-shaped fibrocartilaginous **menisci** that deepen the articular surface, lessen sideways rocking of the femur, and increase shock absorption.

Posteriorly and laterally the femoral and tibial condyles are enclosed by the joint capsule, however the knee is unique in that the articular capsule does not enclose the anterior portion of the joint. Instead, the broad **patellar ligament** extends from the patella to the tibial tuberosity. Because the patellar ligament is located external to the articular capsule, it is classified as an **extracapsular ligament.** Other extracapsular ligaments include the **tibial (medial) collateral ligament**

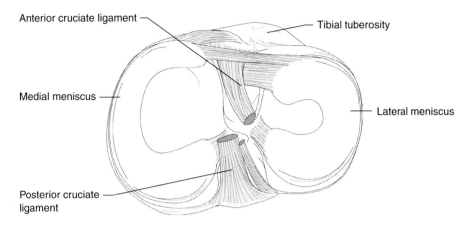

FIGURE 10.5 ◆ Superior View of Menisci and Intracapsular Ligaments on Tibia

and **fibular (lateral) collateral ligament,** both of which help prevent side to side movement of the femur on the tibia. These ligaments are also instrumental in preventing rotation of the knee when it is extended.

Intracapsular ligaments are located within the articular capsule. These include the **anterior cruciate ligament** and **posterior cruciate ligament,** which form an X within the intercondylar notch of the femur. The anterior cruciate ligament originates on the anterior intercondylar region of the tibia and passes posteriorly to insert on the posteromedial aspect of the lateral femoral condyle. When the knee is extended this ligament prevents hyperextension and forward sliding of the tibia. The posterior cruciate ligament originates on the posterior intercondylar region of the tibia and passes anteriorly to insert on the anteromedial aspect of the medial femoral condyle. This ligament prevents forward movement of the femur on the tibia.

Muscle tendons of the hamstring muscles further stabilize the knee joint posteriorly. Hamstring muscles include the biceps femoris, semimembranosus and semitendinosus. Anteriorly, the tendon of the quadriceps inserts into the patella.

◆ Exercise 10.1 *Examine Shoulder, Elbow, and Knee Joint Models*

Using models of the shoulder, rotator cuff, elbow and knee, identify all of the joint structures in bold face in this lab. During your examination, review the primary stabilizing features of each joint. Consider the differences in stability and function between all of the joints you observe.

◆**Exercise 10.2** *Identify Structures in a Cow Knee Joint*

Examine a beef knee joint and use Photo 360c to help identify the following structures:

1. Diaphysis and Epiphyses
2. Epiphyseal line
3. Medullary cavity with yellow marrow
4. Spongy bone in epiphysis
5. Articular cartilage
6. Meniscus
7. Cruciate ligaments

Name _____

Lab 10: Synovial Joints

◆ Practice

1. Choose the movements for each joint listed below from the following options. Some options may be used more than once.

   a. abduction
   b. adduction
   c. circumduction
   d. depression
   e. dorsiflexion
   f. elevation
   g. eversion
   h. extension
   i. flexion
   j. inversion
   k. plantarflexion
   l. pronation
   m. protraction
   n. retraction
   o. rotation
   p. supination

   _____ ankle

   _____ glenohumeral

   _____ elbow

   _____ tibiofemoral

   _____ coxal

   _____ radioulnar

   _____ temperomandibular (TMJ)

2. This diagram depicts a synovial joint. Color the circles next to the names of the joint structures, and then color the corresponding structures in the diagram with the same color.

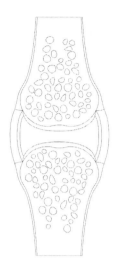

○ Articular cartilage

○ Fibrous capsule

○ Spongy bone

○ Synovial cavity with synovial fluid

○ Synovial membrane

3. The diagrams below show the rotator cuff at the shoulder joint. Color the circles next to the names of the joint structures, and then color the corresponding structures in the diagram with the same color. Draw lines and labels on the diagram for the structures listed with an arrow.

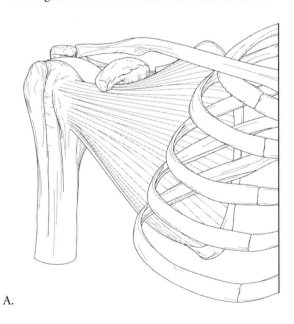

A.

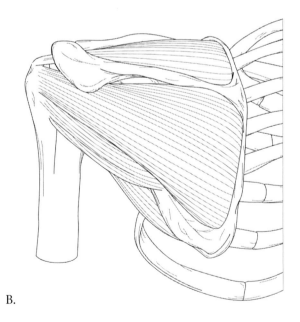

B.

○ Clavicle   ○ Subscapularis

○ Humerus   ○ Supraspinatus

○ Infraspinatus   ○ Teres minor

○ Scapula

→ Acromioclavicular joint   → Head of humerus

→ Acromion process   → Intertubercular groove

→ Coracoid process   → Lesser tubercle

→ Greater tubercle   → Scapular spine

4. The diagram below shows the anterior view of a flexed knee. Draw lines and labels on the diagram for the structures listed below with an arrow.

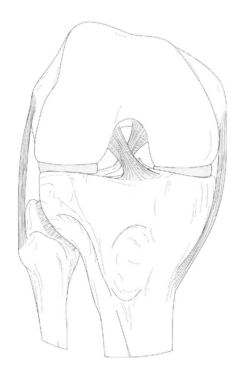

→ Anterior cruciate ligament   → Medial tibial condyle

→ Fibular collateral ligament   → Menisci

→ Intercondylar notch   → Posterior cruciate ligament

→ Lateral femoral condyle   → Tibial collateral ligament

→ Lateral tibial condyle   → Tibial tuberosity

→ Medial femoral condyle   → Tibiofibular joint

5. List the primary stabilizing factors for each of the listed joints.

    a. shoulder

    b. elbow

    c. knee

6. The fibrous capsule is continuous with which membrane that surrounds bone?

7. Explain how each of the following structures serves to stabilize synovial joints.

    a. articular surface

    b. ligaments

    c. muscle tone in tendons

    d. menisci

MUSCLE HISTOLOGY

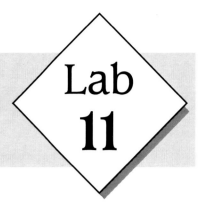

Lab
11

Objectives

1. Describe the location and the function of each muscle tissue type
2. Describe the microscopic structure of skeletal muscle myofilaments
3. Identify smooth, cardiac and skeletal muscle types in the microscopic field of view
4. Identify the structures present at the neuromuscular junction

Materials

- Microscope slides
  - Skeletal muscle
  - Neuromuscular junction
  - Smooth muscle
  - Cardiac muscle

INTRODUCTION

Closely associated with the skeletal system and joints we have studied in the last few labs, is skeletal muscle of the muscular system. Skeletal muscles move the bones and provide posture so that we can move and stand. They also stabilize the joints that they cross. Other types of muscle also function in movement. Smooth muscle squeezes the walls of hollow organs so that substances within them are propelled forward, and cardiac muscle within the walls of the heart moves blood. The property of muscle tissue that contributes to its ability to move other structures is that of *contractility,* or shortening. When a muscle contracts it must convert the energy contained within ATP into mechanical energy. One of the byproducts of contraction is the production of heat. Thus, another function of muscle tissue, particularly skeletal muscle, is the generation of body heat.

CONCEPTS

I. Properties of Muscle Tissue

Although the three muscle types can all contract and are involved in movement, they each have unique locations and functions. Each muscle type has specific structural adaptations related to its locations and functions. Skeletal and cardiac muscle both appear striped, a property we term **striated.** These striations result from the position of tiny subunits called myofilaments within the cells. Cell shape and the position and number of nuclei are different for the individual muscle types as well. A summary of the properties of the different muscle types is presented in Table 11.1.

Table 11.1
Properties of Muscle Types

◆◆◆

Property	Smooth	Cardiac	Skeletal
Cell and Tissue Shape	"squashed football"; form flat sheets of cells; cells in alternating sheets run in perpendicular directions	Short, branched cells in chains; junctions between cells are called **intercalated discs**	Long, parallel unbranched cylindrical fiber; cell is called a **muscle fiber**
Number and Position of Nuclei	Uninucleate, centrally located	Uninucleate or binucleate, generally centrally located	Multinucleate; located peripherally along the length of the muscle fiber
Striations	None	Striated	Striated
Control	Involuntary	Involuntary	Voluntary
Locations	Walls of hollow organs (except heart), including digestive viscera and blood vessels	Walls of the heart	Attached to bones
Functions	Move substances through hollow organs	Move blood through heart	Move bones, provide posture

II. Microscopic Properties of Skeletal Muscle

Skeletal muscle cells bear the name **muscle fibers** because they are long and cylindrical. The plasma membrane of a muscle fiber is specifically termed the **sarcolemma.** Packed tightly within the muscle fiber are long **myofibrils** that run the length of the cell. These structures are composed of thick and thin **myofilaments.** The organization of the myofilaments creates the striated appearance of muscle fibers. Thin filaments are anchored into **Z-discs** that are spaced regularly along the myofilament. Centered between Z-discs, and surrounded by thin filaments, are the thick filaments. A **sarcomere** is the structural unit of skeletal muscle, and is the space between one Z-disc and the next. Upon contraction, the sarcomeres shorten as the thick filaments pull the thin filaments in toward the center of the sarcomere.

The dark striations are known as A bands (remember that there is an "a" in "dark"). The light stripes are called I bands (remember that there is an "i" in "light"). A bands correspond to the region of the sarcomere where thick filaments are present. I bands correspond to that region where no thick filaments are present. In the center of each A band, but too small to view in a light microscope, is a faded region called the **H zone.** This area is the part of the A band that does not contain thin filaments. A dark line down the middle of the H zone is called the **M line,** in which the thick filaments are anchored. The H zone narrows or disappears when the thin filaments meet at the center of the sarcomere during contraction.

Because all of the myofilaments in all of the myofibrils are aligned in a muscle fiber, the fiber contracts as a coordinated whole.

Myofibril

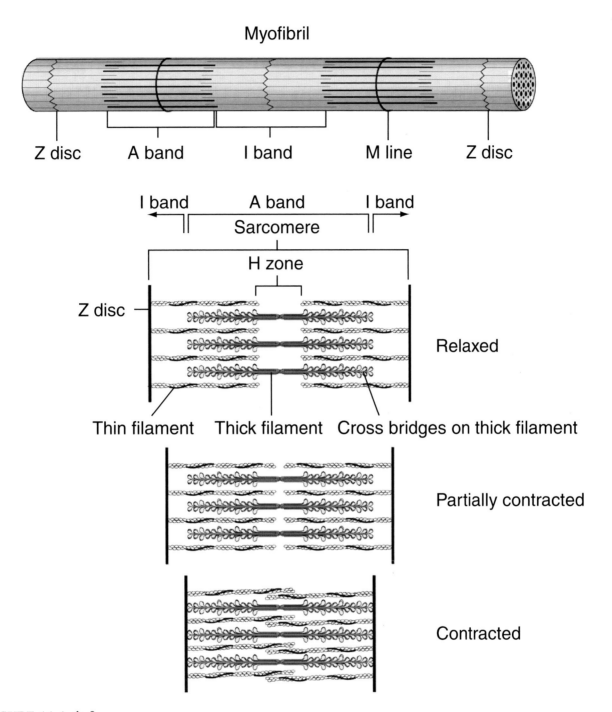

FIGURE 11.1 ◆ Sarcomeres

◆ Exercise 11.1 *Identify Muscle Types and Their Features in the Microscope*

Smooth Muscle

Examine a microscope slide of smooth muscle. Recall that alternating sheets of smooth muscle are arranged with cells running in perpendicular directions. Thus most microscopic sections of smooth muscle will have both longitudinal fibers and cross sections. Refer to Photo 361a as you sketch and label *cells* and their *nuclei* in a longitudinal view in the space below.

Cardiac Muscle

Using Photo 361b as a reference, sketch and label cardiac *cells, intercalated discs, cell branches, striations* and the *nuclei* of cardiac muscle tissue in the space below.

Skeletal Muscle

Obtain a slide showing the longitudinal view of skeletal muscle tissue. Using Photos 361c–362b as a reference, identify and sketch the *muscle fibers, nuclei, sarcolemma, A* and *I bands* in the space below.

III. Neuromuscular Junction

Skeletal muscles receive stimuli for contraction from a type of nerve cell called a **motor neuron.** The **cell body** of the motor neuron is located in the spinal cord, and its **axon** exits the spinal cord to eventually make a connection, or **synapse,** with a muscle. This connection between a motor neuron and a skeletal muscle fiber is called the **neuromuscular junction.** Axonal ends, called axon terminals, do not actually touch the muscle fiber sarcolemma. To communicate with the muscle fiber, the axon terminal releases a chemical, **acetylcholine** (ACh), into the shallow **synaptic cleft.** ACh diffuses across the synaptic cleft to bind to receptors embedded in a folded region of sarcolemma called the **motor end plate.** Receptors that have bound ACh open ion channels and allow sodium to enter the muscle fiber cytosol, depolarizing the motor end plate and initiating the steps of contraction. Potassium also exits the muscle cell through the open ion channels.

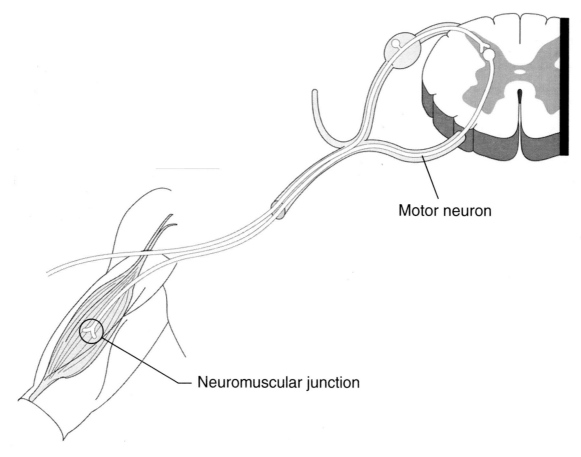

FIGURE 11.2 ◆ Neuromuscular Junction

◆**Exercise 11.2** *Examine a Microscope Slide of the Neuromuscular Junction*

Obtain a microscope slide with a neuromuscular junction. Observe the specimen using the 10X and the 40X objective. The spotted oval region in the muscle fibers at the end of the axons is the motor end plate. You will see several axons traveling together as a nerve, from which individual axons exit to form neuromuscular junctions with muscle fibers. One axon may branch and make connections with multiple muscle fibers. Using Photo 362c as a reference, sketch the neuromuscular junction and label the *muscle fiber, axons, nerve* and *motor end plate* in the space below.

Name _____

Lab 11: Muscle Histology

◆ Practice

1. The three types of muscle tissues exhibit similarities as well as differences. Check the appropriate boxes in the table below to indicate which muscle type(s) exhibits each characteristic.

Characteristic	Skeletal	Cardiac	Smooth
Voluntarily controlled	✓		
Involuntarily controlled		✓	✓
Striated	✓	✓	
Has a single nucleus (or 2) in each cell		✓	
Has several nucleus per cell	✓		
Found attached to bones	✓		
Allows you to move your eyeballs	✓		
Found in the walls of the stomach, uterus & arteries			✓
Contains cells shaped like squashed footballs			✓
Contains short branching, cylindrical cells		✓	
Contains long, nonbranching cylindrical cells	✓		
Has intercalated discs		✓	
Concerned with locomotion of the body as a whole	✓		
Changes the internal volume of an organ as it contracts		✓	✓
Located in the heart wall		✓	

2. Label the regions of the sarcomere in the diagram using the list below.

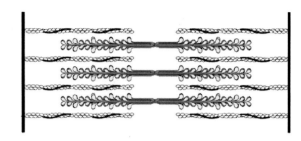

→ A band

→ H zone

→ I band

→ M line

→ Sarcomere

→ Thick filament

→ Thin filament

3. For each of the muscle types, list their location(s) and functions:

a. Smooth muscle
Location:

Functions:

b. Cardiac muscle
Location:

Functions:

c. Skeletal muscle
Location:

Functions:

4. For each of the muscle types, what features would you observe in microscopic analysis?

    a. Smooth

    b. Cardiac

    c. Skeletal

MUSCULATURE

Objectives

1. Define **agonist, antagonist, synergist, origin** and **insertion**
2. Name and locate the major muscles of the human body on models and diagrams, and state the action of each
3. Name muscle origins and insertions *as required by the instructor*
4. Practice basic dissecting techniques
5. Name and locate the major muscles in preserved cats

Materials

- Human Muscle Models
  - Torso
  - Leg
  - Arm
- Preserved cats
- Dissecting trays
- Disposable gloves
- Dissecting instruments

INTRODUCTION

More than 600 muscles are present in the human body, ranging in size from the tiny stapedius muscle in the middle ear to the large rectus femoris muscle of the quadriceps. As with organs in other systems, muscles can be classified according to different organizational schemes. Muscle size, direction of muscle fibers, muscle shape, and number of origins are all methods of describing various muscles. In this lab we will group muscles according to their actions at certain regions of the body.

Muscle action is determined by the attachment of the muscle to bones on either side of a joint. The attachment of a muscle to the less moveable bone of a joint is called the **origin,** while the attachment to the more moveable bone is termed the **insertion.** Some muscles have multiple origins. Muscle attachments and location determine the muscle's actions. The primary mover of a specific joint (e.g. elbow flexor or neck extender) is called the **agonist. Synergists** are muscles that assist the agonist by stabilizing bones to reduce unnecessary movement. For example, rhomboideus muscles stabilize the scapula during movements of the humerus. Agonistic movements are opposed or reversed by **antagonists.** The triceps brachii muscle reverses elbow flexion caused by the biceps brachii. This means that, for elbow extension, the triceps is the agonist and the biceps is an antagonist.

CONCEPTS

I. Muscle Names

Muscle names give descriptive clues that can help you remember their locations, actions or characteristics. As you identify the major muscles of the human body, learn to recognize the characteristic for which each muscle is named.

Table 12.1
Muscle Naming Conventions

- **Direction of muscle fibers**
  - Rectus—fibers run parallel to the midline of the torso
  - Transversus—fibers run perpendicularly to the midline of the torso
  - Oblique—fibers run at an angle to the midline of the torso
- **Size of muscle**
  - Maximus—largest muscle of that name
  - Minimus—smallest muscle of that name
  - Longus—longest muscle of that name
  - Brevis—shortest muscle of that name
- **Location of muscle**
  - Some muscles are named for their origin or location on a specific bone
- **Number of origins**
  - Biceps—muscle has two points of origin
  - Triceps—muscle has three points of origin
  - Quadriceps—muscle has four points of origin
- **Shape of muscle**
  - Deltoid—triangular
  - Trapezius—trapezoidal
  - Serratus—serrated
- **Action of muscle**
  - Some muscles are named for the action that they perform

II. Identification of Human Muscles

Table 12.2
Muscles of Facial Expression (Photo 363a)

Muscle Name	Figure	Origin	Insertion	Action
Platysma	12.1	Fascia covering pectoral and deltoid	Mandible, skin of chin and cheek	Depresses mandible and lower lip, tenses skin of neck
Frontalis	12.1	Cranial aponeurosis over frontal bone	Skin of eyebrow & bridge of nose	Raises eyebrow, wrinkles forehead, draws scalp anteriorly
Orbicularis oris	12.1	Multiple sites of origin from the maxillae and mandible	Muscle and skin surrounding the mouth	Purse and protrude lips, as in kissing
Orbicularis oculi	12.1	Frontal and maxillary bones	Eyelid	Blinking and squinting movements of eyelid

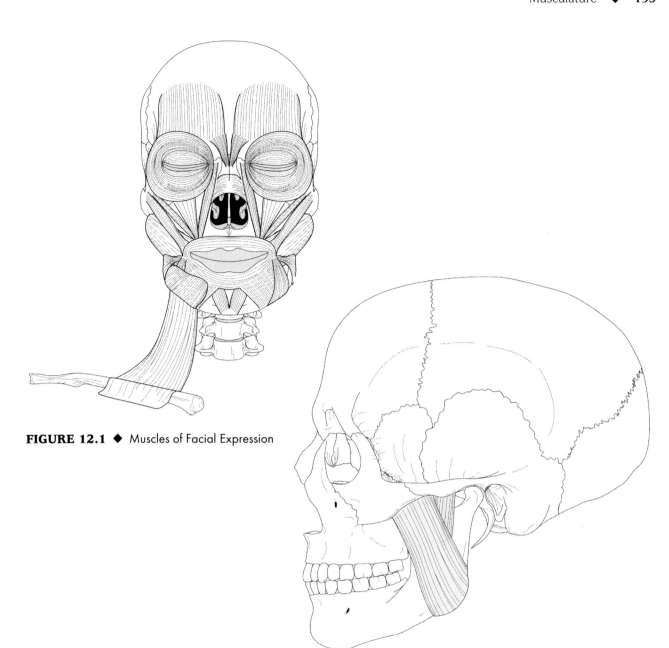

FIGURE 12.1 ◆ Muscles of Facial Expression

FIGURE 12.2 ◆ Masseter

Table 12.3
Muscles of Mastication (Photo 363a)

◆◆◆

Muscle Name	Figure	Origin	Insertion	Action
Masseter	12.2	Zygomatic arch and maxilla	Angle and ramus of mandible	Elevates and protracts mandible
Temporalis	12.3	Temporal lines on temporal and frontal bones	Coronoid process and ramus of mandible	Elevates and retracts mandible

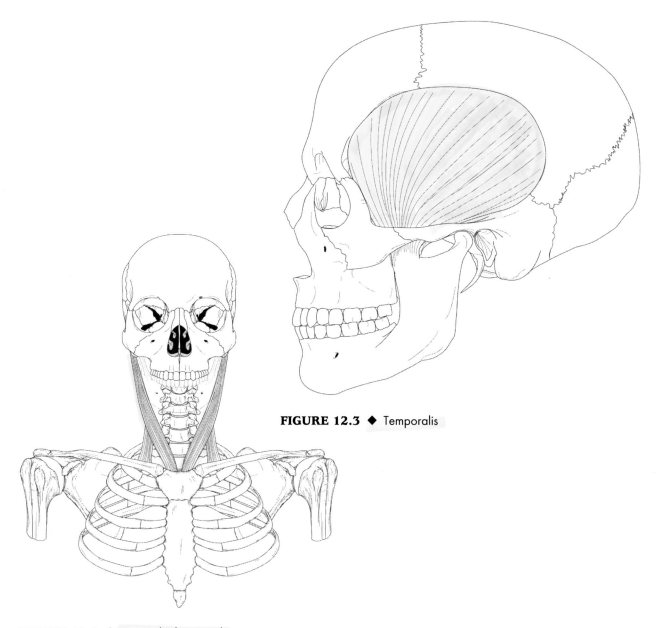

FIGURE 12.3 ◆ Temporalis

FIGURE 12.4 ◆ Sternocleidomastoid

Table 12.4
Muscles That Move the Head (Photo 363b)

◆◆◆

Muscle Name	Figure	Origin	Insertion	Action
Sternocleidomastoid	12.4	Manubrium of sternum and medial clavicle	Mastoid process of temporal bone	Individually medially rotate neck, together they flex the neck
Trapezius	12.7	Occipital bone, ligamentum nuchae, spines of C_7–T_{12}	Lateral clavicle, acromion and spine of scapula	Extends neck, elevates, rotates and retracts scapula

Table 12.5

Muscles That Move the Shoulder (Photo 363b)

◆◆◆

Muscle Name	Figure	Origin	Insertion	Action
Serratus anterior	12.5	Lateral margin of ribs 1–8	Vertebral border of anterior scapula	Rotates scapula superiorly and laterally, protracts scapula
Pectoralis minor	12.6	Medial margin of ribs 3–5	Coracoid process of scapula	Depresses and protracts scapula, elevates ribs 3–5 when scapula is fixed
Rhomboideus minor and major	12.7	Spinous processes of C_7–T_5	Vertebral border of scapula	Retracts scapula and rotates it to tilt glenoid cavity downward
Trapezius	12.7	Occipital bone, ligamentum nuchae, spines of C_7–T_{12}	Lateral clavicle, acromion and spine of scapula	Extends neck, elevates, rotates and retracts scapula

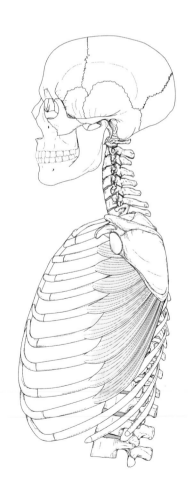

FIGURE 12.5 ◆ Serratus Anterior

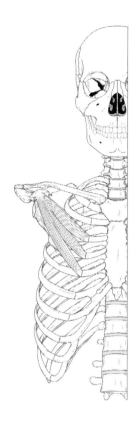

FIGURE 12.6 ◆ Pectoralis Minor

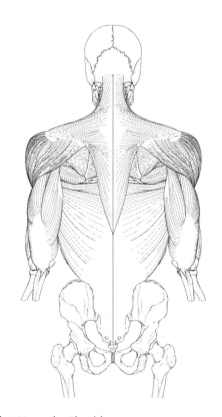

FIGURE 12.7 ◆ Posterior Muscles That Move the Shoulder

Table 12.6
Muscles That Move the Arm (Photos 363b—364b)

◆◆◆

Muscle Name	Figure	Origin	Insertion	Action
Pectoralis major	12.8	Clavicle, sternum, costal cartilages of ribs 1–6	Greater tubercle of humerus	Flexes, adducts and medially rotates arm
Latissimus dorsi	12.7	Spinous processes of T_7 to L_5 and the sacrum, posterior iliac crest, ribs 10–12	Intertubercular groove of humerus	Extends, adducts and medially rotates arm
Deltoid	12.7, 12.8	Clavicle, acromion and spine of scapula	Deltoid tuberosity of humerus	Abducts arm, assists with medial and lateral rotation of arm
Supraspinatus*	12.9	Supraspinous fossa	Greater tubercle of humerus	Abducts arm
Infraspinatus*	12.9	Infraspinous fossa	Greater tubercle of humerus	Laterally rotates and adducts arm
Subscapularis*	12.10	Subscapular fossa	Lesser tubercle of humerus	Medially rotates arm
Teres minor*	12.9	Lateral margin of scapula	Greater tubercle of humerus	Laterally rotates and adducts arm
Teres major	12.9	inferior angle of scapula	Intertubercular groove of humerus	Extends, medially rotates, and adducts arm

*rotator cuff muscles

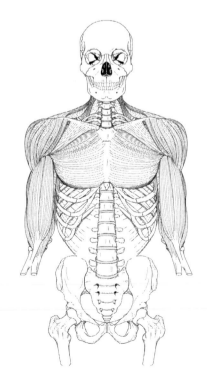

FIGURE 12.8 ◆ Anterior Muscles That Move the Shoulder

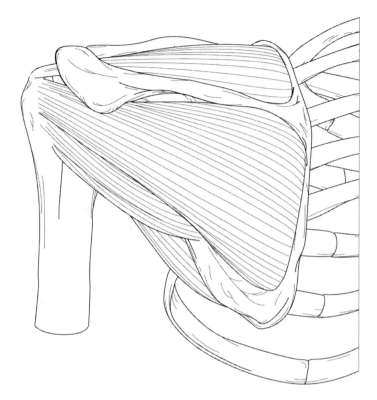

FIGURE 12.9 ◆ Posterior Rotator Cuff Muscles and Teres Major

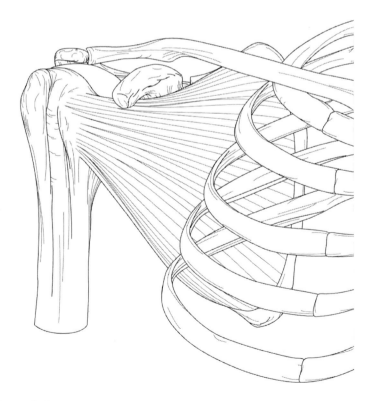

FIGURE 12.10 ◆ Subscapularis

Table 12.7
Muscles That Move the Forearm (Photos 364a—364b)

◆◆◆

Muscle Name	Figure	Origin	Insertion	Action
Biceps brachii*	12.8, 12.11	One origin above glenoid fossa, another on the coracoid process of scapula	Radial tuberosity	Flexes and supinates forearm, flexes arm
Brachialis	12.11	Distal anterior surface of humerus	Coronoid process of ulna	Flexes forearm
Triceps brachii*	12.7, 12.12	Long head arises below glenoid fossa, lateral head arises from lateral and posterior proximal shaft of humerus proximally, medial head arises from proximal posterior surface of humerus	Olecranon process of ulna	Extends forearm

* These two muscles cross both shoulder and elbow joints; therefore, they also move the arm.

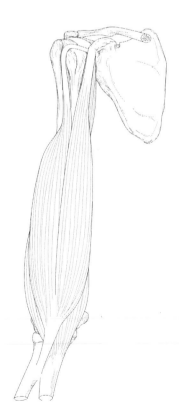

FIGURE 12.11 ◆ Anterior Brachial Muscles

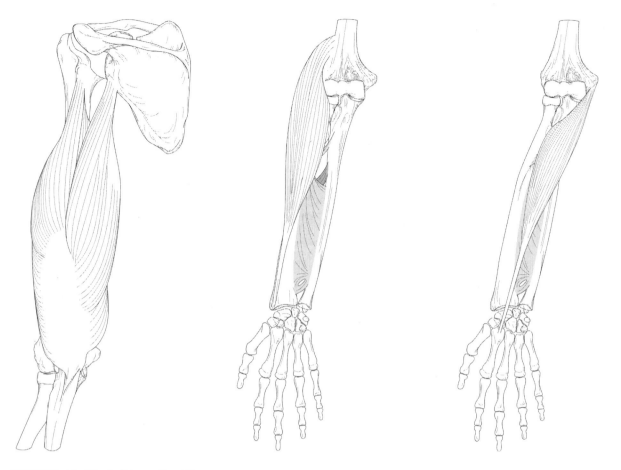

FIGURE 12.12 ◆ Triceps Brachii   **FIGURE 12.13** ◆ Brachioradialis   **FIGURE 12.14** ◆ Flexor Carpi Radialis

Table 12.8
Muscles That Move the Forearm, Hand, and Fingers

◆◆◆

Muscle Name	Figure	Origin	Insertion	Action
Anterior Compartment of Forearm (Photo 364b)				
Brachioradialis	12.13, 12.17	Lateral borders of distal end of humerus	Styloid process of radius	Flexes and supinates forearm
Flexor carpi radialis	12.14, 12.17	Medial epicondyle of humerus	2nd and 3rd metacarpals	Flexes and abducts wrist
Palmaris longus	12.15, 12.17	Medial epicondyle of humerus	Palmar fascia	Flexes wrist
Flexor carpi ulnaris	12.16, 12.17, 12.18	Medial epicondyle of humerus, posterior olecranon process of ulna	5th metacarpal and pisiform bone	Flexes and adducts hand
Posterior Compartment of Forearm (Photo 364a)				
Extensor carpi radialis longus	12.18	Lateral epicondyle of humerus	2nd metacarpal	Extends and abducts hand
Extensor carpi radialis brevis	12.18	Lateral epicondyle of humerus	3rd metacarpal	Extends and abducts hand
Extensor digitorum	12.18	Lateral epicondyle of humerus	2nd–5th middle and distal phalanges	Extends phalanges and wrist
Extensor carpi ulnaris	12.18	Lateral epicondyle of humerus, posterior border of ulna	5th metacarpal	Extends and adducts hand

FIGURE 12.15 ◆ Palmaris Longus

FIGURE 12.16 ◆ Flexor Carpi Ulnaris

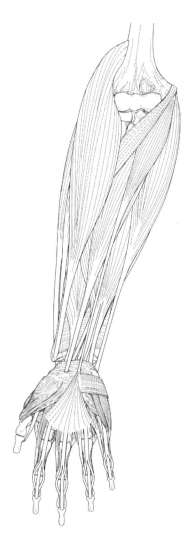

FIGURE 12.17 ◆ Anterior Antebrachial Muscles

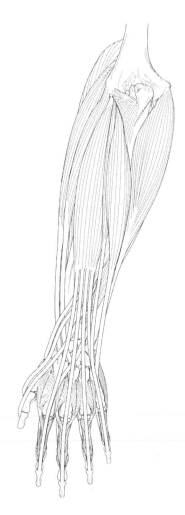

FIGURE 12.18 ◆ Posterior Antebrachial Muscles

Table 12.9
Muscles That Move the Thigh and Leg

◆◆◆

Muscle Name	Figure	Origin	Insertion	Action
Anterior Compartment of the Thigh (Photos 365a and 365b)				
Rectus femoris*	12.19, 12.24	Anterior inferior iliac spine	Tibial tuberosity via patellar ligament	Extends leg, flexes thigh
Vastus lateralis*	12.20, 12.24	greater trochanter and linea aspera of femur	Tibia tuberosity via patellar ligament	Extends leg
Vastus medialis*	12.21, 12.24	Linea aspera of femur	Tibial tuberosity via patellar ligament	Extends leg
Vastus intermedius*	12.22	Anterior surface of femur	Tibial tuberosity via patellar ligament	Extends leg
Sartorius	12.23, 12.24	Anterior superior iliac spine	Medial surface of tibia near tibia tuberosity	Flexes and laterally rotates thigh, flexes leg

* quadriceps muscles

Medial Compartment of the Thigh (Photo 365b)				
Pectineus	12.25, 12.29	Superior ramus of pubis	Below lesser trochanter on posterior aspect of femur	Adducts, flexes and medially rotates thigh
Adductor magnus	12.26, 12.29	Inferior ramus of pubis	Linea aspera of femur	Adducts, flexes and medially rotates thigh
Adductor longus	12.27, 12.29	Inferior ramus of pubis	Linea aspera of femur	Adducts, flexes and medially rotates thigh
Gracilis	12.28, 12.29	Pubis below symphysis	Medial surface of tibia inferior to medial condyle	Adducts, flexes and medially rotates thigh

Lateral Compartment of the Thigh (Photo 365a)				
Tensor fasciae latae	12.30	Iliac crest	Tibia via iliotibial tract	Abducts thigh

Posterior Compartment of the Thigh (Photos 365a and 365b)				
Gluteus maximus	12.31	Iliac crest, sacrum, coccyx	Gluteal tuberosity of femur and iliotibial tract	Extends and laterally rotates thigh
Gluteus medius	12.31	Lateral surface of ilium	Greater trochanter of femur	Extends and laterally rotates thigh
Biceps femoris*	12.32	Long head arises from ischial tuberosity, short head arises from linea aspera of femur	Lateral condyle of tibia and head of fibula	Flexes leg, extends thigh
Semitendinosus*	12.32	Ischial tuberosity	Medial surface of proximal tibia	Flexes leg, extends thigh
Semimembranosus*	12.32	Ischial tuberosity	Medial condyle of tibia	Flexes leg and extends thigh

* hamstrings

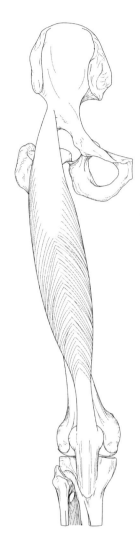

FIGURE 12.19 ◆ Rectus Femoris

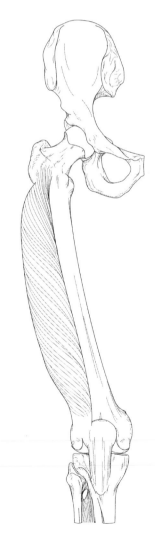

FIGURE 12.20 ◆ Vastus Lateralis

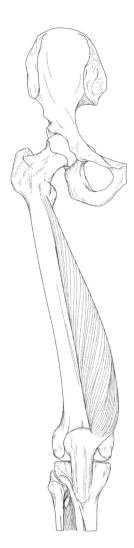

FIGURE 12.21 ◆ Vastus Medialis

FIGURE 12.22 ◆ Vastus Intermedius

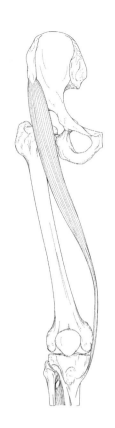

FIGURE 12.23 ◆ Sartorius

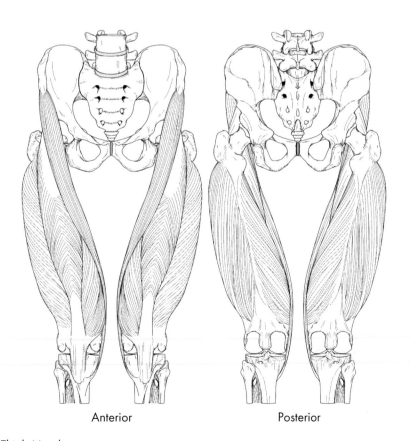

Anterior   Posterior

FIGURE 12.24 ◆ Thigh Muscles

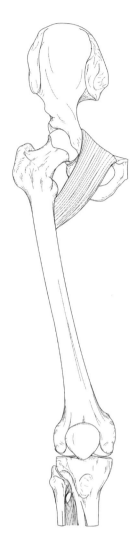

FIGURE 12.25 ◆ Pectineus

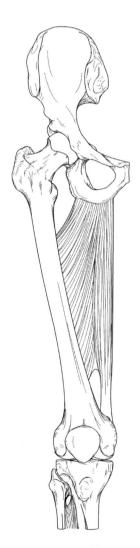

FIGURE 12.26 ◆ Adductor Magnus

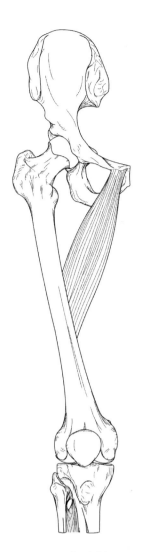

FIGURE 12.27 ◆ Adductor Longus

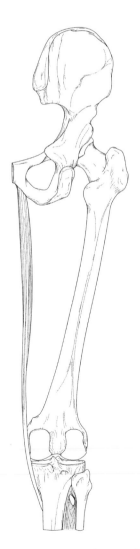

FIGURE 12.28 ◆ Gracilis

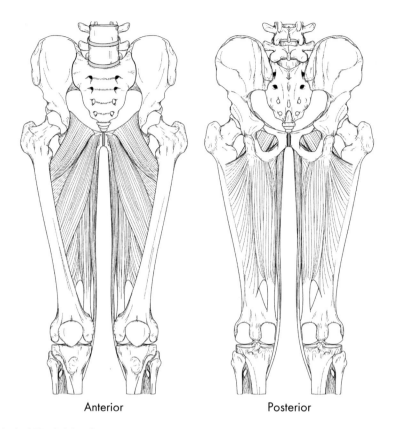

Anterior   Posterior

FIGURE 12.29 ◆ Medial Thigh Muscles

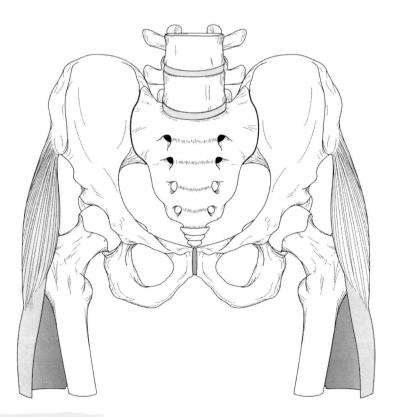

FIGURE 12.30 ◆ Tensor Fasciae Latae

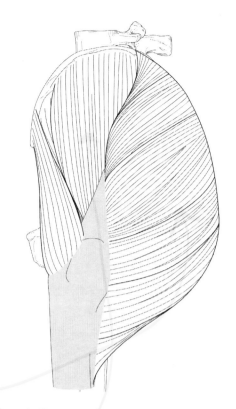

FIGURE 12.31 ◆ Gluteus Maximus and Gluteus Medius

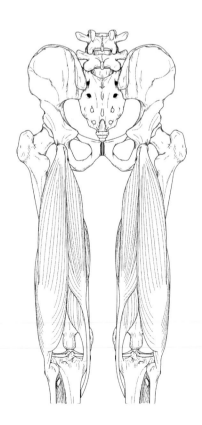

FIGURE 12.32 ◆ Hamstring Muscles

Table 12.10
Muscles That Move the Foot and Toes

◆◆◆

Muscle Name	Figure	Origin	Insertion	Action
Anterior Compartment of the Leg (Photos 365a and 365b)				
Tibialis anterior	12.33	Lateral condyle of tibia, interosseous membrane	1st metatarsal, medial cuneiform	Dorsiflexes, inverts foot
Extensor digitorum longus	12.34	Lateral condyle of tibia, proximal fibula, interosseous membrane	2nd–5th middle and distal phalanges	Dorsiflexes and everts foot, extends toes
Lateral Compartment of the Leg (Photo 365a)				
Peroneus longus	12.35	Lateral condyle of tibia, proximal half of fibula	1st metatarsal, medial cuneiform	plantar flexes, everts foot
Posterior Compartment of the Leg (Photos 365a and 365b)				
Gastrocnemius	12.36	Lateral and medial condyles of femur	Calcaneus via Achilles (calcaneal) tendon	Plantar flexes foot, flexes leg
Soleus	12.36	Proximal fibula and tibia	Calcaneus via Achilles (calcaneal) tendon	Plantar flexes foot

FIGURE 12.33 ◆ Tibialis Anterior

FIGURE 12.34 ◆ Extensor Digitorum Longus

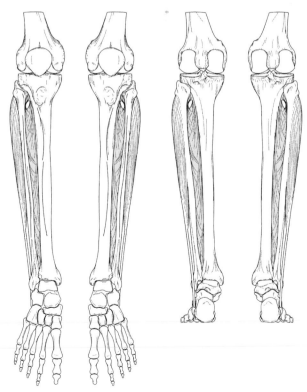

Anterior   Posterior

FIGURE 12.35 ◆ Peroneus Longus

FIGURE 12.36 ◆ Posterior Crural Muscles

Table 12.11
Muscles of the Abdominal Wall (Photo 363b)

◆◆◆

Muscle Name	Figure	Origin	Insertion	Action
External abdominal oblique	12.37	Lower 8 ribs	Linea alba & iliac crest	Compresses abdomen, lateral rotation of vertebral column
Rectus abdominis	12.38	Superior surface of pubis around symphysis	Xiphoid process of sternum and costal cartilages of ribs 5–7	Compresses abdomen, flexes vertebral column
Internal oblique	12.39	Iliac crest and lumbodorsal fascia	Costal cartilages of lower 3 ribs, linea alba and pubic crest	Compresses abdomen, medial rotation of vertebral column
Transversus abdominis	12.38	Lower 6 ribs, iliac crest and lumbodorsal fascia	Pubis, linea alba and costal cartilages of ribs 5–7	Compresses abdomen

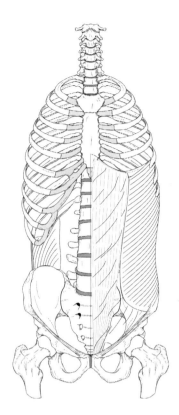

FIGURE 12.37 ◆ External Abdominal Oblique

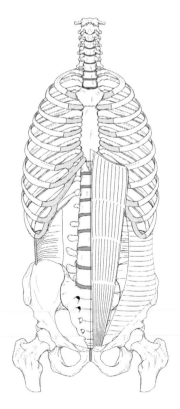

FIGURE 12.38 ◆ Rectus Abdominis and Transversus Abdominis

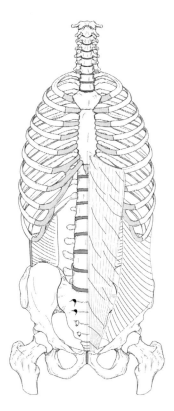

FIGURE 12.39 ◆ Internal Abdominal Oblique

Table 12.12
Muscles Used in Breathing

Muscle Name	Figure	Origin	Insertion	Action
Diaphragm	12.40	Xiphoid process, costal cartilages of lower 6 ribs and lumbar vertebrae	Central tendon of diaphragm	Increases size of thoracic cavity for inspiration
External intercostals	12.41	Inferior border of each rib	Superior border of next rib inferiorly; fibers angled anteriorly	elevates ribs in quiet inspiration
Internal intercostals	12.41	Superior border of each rib	Inferior border of the next rib superiorly; fibers angled posteriorly	Depresses ribs in forceful expiration
Scalenes	12.41	Transverse processes of cervical vertebrae	Anterolateral aspect of ribs 1 and 2	Elevates first two ribs in deep inspiration
Sternocleidomastoid	12.4	Manubrium of sternum and medial clavicle	Mastoid process of temporal bone	Elevates the clavicle and sternum in deep inspiration

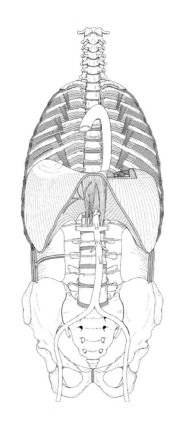

FIGURE 12.40 ◆ Diaphragm

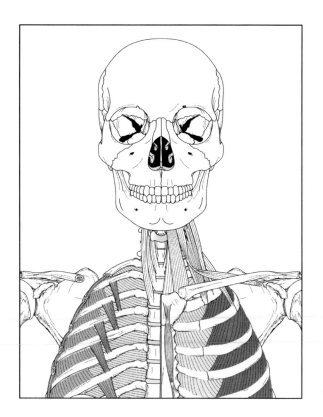

FIGURE 12.41 ◆ External Intercostals, Internal Intercostals, and Scalenes

◆ Exercise 12.1   *Label Muscle Diagrams*

Label and color all muscles from Lab 12 tables in Figures 12.1–12.41. Review actions of the muscles as you label them.

◆◆◆

◆ Exercise 12.2   *Identify Muscles on Human Muscle Models*

Identify each of the muscles from Lab 12 tables on a torso, leg, and arm muscle model. Review actions of the muscles as you identify them.

◆◆◆

III. Dissection and Identification of Cat Muscles

Dissection of cat muscles will further increase your understanding of muscle position, attachment and texture. As you progress through the exercise, take time to palpate significant attachments and bony landmarks. Evaluate the muscle's action based on its location across joints. Ask yourself how the muscle action in the cat might differ from that of the same muscle in a human.

The cat specimens have been preserved and prepared for student exploration. Skin has been previously removed, and the arteries and veins have been injected with red and blue latex for ease of identification. Keep in mind that directional terminology is different when describing quadruped animals. However we are using the cat as a model for *human* anatomy and so will apply human directional terms to the cat to avoid confusion.

Please be aware that your cat will be used for future dissection exercises. **Do not dissect any structures that are not described in this lab.** It is especially important to maintain the integrity of the internal body cavities for future exploration.

◆ Exercise 12.3   *Dissect and Identify Muscles on a Preserved Cat*

Wear disposable gloves during the dissection exercise. Your most useful dissection instruments will likely be a blunt probe, scissors, forceps and your fingers. Remember not to point to structures with a sharp instrument — someone else may be pointing at the same structures with their fingers. Rotate "dissection duty" among your lab partners so that everyone can share the experience. The lab partners who are not currently dissecting can help by reading the lab exercise instructions and finding the appropriate photos from the Photo Atlas in the back of this lab manual.

A. Anterior Thorax Region (Photos 400, 401a, and 404)

Lay your cat on its back on the dissecting tray. You may need to remove superficial fatty tissue and some fascia from the surface of the cat's muscles. Be careful not to gouge the muscle tissue as you clean the surface of the muscles.

Begin by identifying the **pectoralis major** and **pectoralis minor** muscles on the left side. In the cat the pectoralis major is the smaller muscle, shaped like a small triangle at the superior aspect of the thorax. Pectoralis minor is a larger triangle, the superior margin of which is located deep to pectoralis major. Carefully define the superior and inferior margins of pectoralis major, and the inferior margin of pectoralis minor.

Next, on left side, identify and define the anterior margin of the **latissimus dorsi** muscle extending from the lateral aspect of the cat to its back. Palpate the muscle's attachment to the vertebral spines. Note that the anterior margin continues through the axillary region.

On the *right* side of the anterior thorax, cut the latissimus dorsi in the axillary region and reflect the muscle. Be careful to leave the muscle attached for reference. Deep to the latissimus dorsi, attached to the rib cage, is the fanshaped **serratus anterior** muscle. In a quadruped, this muscle is called the serratus ventralis. Observe the origins on the individual ribs. Carefully separate the serrated points of origin to better observe the unique shape of this muscle.

B. Abdominal Region (Photos 400, 401b, and 404)

Observe the angled fibers of the **external oblique** muscles and identify their insertion into the broad aponeurosis. Carefully cut a shallow window in one side of the aponeurosis to see the underlying fibers of the **rectus abdominis** running parallel with the torso. Try to identify the linea alba at the junction of the two rectus abdominis muscles.

Next, use forceps to gently lift the external oblique at the junction with the aponeurosis. Use the point of your scissors to poke a small, shallow hole in the junction on one side. Close your scissors and insert them into the hole to separate the very thin external oblique from the underlying internal oblique. Once you are certain you can discern the two muscle layers, use the scissors to cut the external oblique along its insertion to the aponeurosis. Make this cut about three to four inches long. Then make lateral incisions in the external oblique to create a flap, which you can reflect.

Deep to the external oblique you will see the **internal oblique** muscle extending from the lateral aspect to insert in another, deeper aponeurosis. Deep to this inner aponeurosis is the **transversus abdominis** muscle. You can expose the fibers of the transverses abdominis by cutting a small window in the inner aponeurosis. Be extremely careful not to cut too deeply and expose the abdominal cavity.

C. Posterior Thorax Region (Photos 399, 402, 403a, 403b, and 405b)

Lay the cat on its anterior surface to expose its back. The **trapezius** muscles form a diamond shape that extends from the cervical region to the lower thoracic region. In the human, the trapezius is a single large muscle on each side of the vertebral column. The cat, however, has three trapezius muscles on each side: the clavotrapezius, acromiotrapezius and spinotrapezius. The prefix of each cat trapezius muscle indicates its insertion point. Define the lateral margin of the spinotrapezius and acromiotrapezius muscles on the left side of the cat. On the right side of the cat, use scissors to make a medial cut through the spino- and acromiotrapezius muscles along the vertebral column. Be careful not to cut into the underlying musculature. At the level of the scapula make a perpendicular incision through the trapezius from the vertebral column to the lateral margin of the muscle. Palpate the attachment of the trapezius muscle to the spine of the scapula. Cut the trapezius away from the spine of the scapula. You should now be able to reflect two flaps of the trapezius to expose the deeper scapular musculature.

The **rhomboideus** muscles connect the inferior angle of the scapula to the vertebral column. Gently pull on the rhomboideus muscles to approximate their action. Once again identify the spine of the scapula. Observe the **supraspinatus** and **infraspinatus** muscles in the supraspinous and infraspinous fossas respectively. The **teres major** muscle passes medially to the humerus from the inferior angle of the scapula, and inserts in the head of the humerus. Review the position of the serratus anterior, and its connection to the scapula. On the anterior aspect of the scapula you will find the **subscapularis** muscle in the subscapular fossa.

D. Arm Region (Photo 403a)

Now lay the cat on one side. As with the trapezius group, the **deltoid** group in the cat is composed of three muscles. The clavodeltoid originates on the clavicle and continues to the forearm. Define the medial and lateral edges of the clavodeltoid and gently move it anteriorly to observe the laterally positioned **brachialis** muscle deep to the clavodeltoid.

The **triceps brachii** muscle is located laterally (superficial lateral head and deep medial head) and posteriorly (long head) on the humerus.

E. Posterolateral Lower Limb (Photos 402, 406a, 406b, and 407)

With the cat on its side, remove the fascia and fatty tissue from the surface of the lateral lower limb. At the anterior aspect, the **sartorius** muscle is visible as a thin strip as it crosses the hip joint laterally but widens into a broad, thin muscle on the medial surface of the lower limb. Posterior to the sartorius muscle and positioned close to the hip is the small, roughly triangular **tensor fasciae latae** muscle. This muscle inserts into a long aponeurosis called the fascia latae, which inserts into the tibia. Use scissors to cut the anterior and posterior edges of the fasciae latae and gently lift it to examine the deeper **vastus lateralis** muscle of the quadriceps group.

Posterior to the fascia latae on the lateral surface is the broad, flat **biceps femoris** muscle, which originates on the ischial tuberosity and inserts on the tibia. In the human, the biceps femoris is located more posteriorly. On the narrow posterior aspect of the femur is positioned the **semitendinosus** muscle that flexes the hip and knee. Deep to the semitendinosus and biceps femoris is the broad **semimembranosus**. Together, the biceps femoris, semi-tendinosus and semimembranosus are called the hamstrings.

The major calf muscles are the **gastrocnemius** and deeper **soleus**. Both insert into the calcaneus via the tough calcaneal (Achilles) tendon. Separate these two muscles for easier identification. Laterally, identify the **peroneus longus** (also called the **fibularis longus**), extensor digitorum longus and **tibialis anterior** muscles. Trace the origins and insertions of these muscles to project their respective eversion and dorsiflexion actions.

F. Medial Lower Limb (Photos 408a and 408b)

Superficially, the anterior **sartorius** and posterior **gracilis** muscles are separated by the femoral artery and vein. Transect and reflect the gracilis muscle to examine the deeper adductor muscles and the semimembranosus. The **adductor femoris** muscle in the cat is homologous to the human **adductor magnus** and **adductor brevis** muscles. Superior to the adductor femoris is the thin **adductor longus** muscle.

G. Review the following muscles in the cat:

1. **Pectoralis major**
2. **Pectoralis minor**
3. **Rectus abdominis**
4. **External oblique**
5. **Internal oblique**
6. **Transversus abdominis**
7. **Serratus anterior** (also called Serratus ventralis in cat)
8. **Deltoid** (identify cat clavodeltoid)
9. **Trapezius** (in cat trapezius has 3 parts: clavotrapezius, acromiotrapezius, spinotrapezius)
10. **Latissimus dorsi**
11. **Rhomboideus**
12. **Supraspinatus**
13. **Infraspinatus**
14. **Subscapularis**
15. **Teres major**
16. **Brachialis**
17. **Triceps brachii** (lateral head, long head)
18. **Tensor fasciae latae**
19. **Biceps femoris**
20. **Semimembranosus**
21. **Semitendinosus**
22. **Sartorius**
23. **Vastus lateralis**
24. **Gracilis**
25. **Tibialis anterior**
26. **Extensor digitorum longus**
27. **Peroneus longus** (also called **fibularis longus**)
28. **Gastrocnemius**
29. **Soleus**
30. **Adductor magnus** (in cat this is part of **Adductor Femoris**)
31. **Adductor brevis** (in cat this is part of **Adductor Femoris**)
32. **Adductor longus**

Name _____

Lab 12: Musculature

◆ **Practice**

1. Fill in the action for each muscle listed below.

Muscle Name	Action
Semimembranosus	
Rectus femoris	
Deltoid	
External oblique	
Flexor carpi radialis	
Latissimus dorsi	
Platysma	
Extensor carpi ulnaris	
Orbicularis oculi	
Palmaris longus	
Gastrocnemius	
Pectoralis minor	
Tibialis anterior	
Masseter	
Sartorius	
Sternocleidomastoid	
Trapezius	

2. For each movement list the agonist and antagonist.

Movement	Agonist	Antagonist
Flexion at elbow		
Extension of the wrist		
Abduction of the arm		
Extension of the neck		
Flexion of the knee		
Dorsiflexion of foot		
Adduction of the thigh		
Abduction of the wrist		
Flexion at the hip		

NERVOUS TISSUE

Lab 13

Objectives

1. Identify the parts of a neuron on diagrams and a model, and know their functions
2. Identify the important anatomical characteristics of neurons on slides
3. Examine three different types of neurons from different regions of the nervous system

Materials

- Model of neuron
- Microscope slides
  - Pyramidal cells from the cerebrum
  - Purkinje cells from the cerebellum
  - Giant Multipolar neurons from the spinal cord

INTRODUCTION

Our bodies are able to react to external and internal stimuli due to activity of the nervous system. We also have the ability to voluntarily control skeletal muscles, and so manipulate certain aspects of our environment. Brain functions such as thinking, emotions and memory are all related to nervous system operations. Indeed, the nervous system is one of the most versatile organ systems, and is interrelated with every other organ system in the body.

Two classes of cells make up nervous tissue. **Supporting cells,** or **glia** (neuroglia), contribute to the physical and environmental health of nerve cells. **Neurons** are electrically active cells that conduct impulses called **action potentials** along cellular extensions. All neurons have the same basic function, however the structure of the cells and the number of cytoplasmic extensions differs depending on location within the nervous system. In this lab you will examine three types of motor neurons from different regions of the brain and spinal cord.

CONCEPTS

I. Structure of a Neuron

The parts of a neuron reflect its function in the generation and transmission of action potentials.

Table 13.1
Parts of a Neuron

◆◆◆

Soma	Also known as the cell body; contains the nucleus; site of protein synthesis
Nucleus	Usually large and centered in the soma
Dendrites	The receptive portion of the cell, detects stimuli and conducts action potentials toward the cell body; number of dendrites varies from one to many
Axon	A single extension from the cell that conducts an action potential away from the soma
Axon hillock	The junction between the soma and axon; site of action potential generation
Nissl bodies	Specialized endoplasmic reticulum found in cell body and cellular extensions
Myelin sheath	Fatty substance secreted from Schwann cells located along the axon; myelin electrically insulates the axon and increases velocity of action potential propagation along axon
Nodes of Ranvier	Gaps in between Schwann cells where no myelin covers the axon
Telodendria	Small branches from the distal end of the axon
Axon terminals	Bulbous terminal ends of the telodendria which contain neurotransmitter; secretory portion of the neuron releases neurotransmitter in response to action potentials
Synapse	Junction between the axon of one neuron and the dendrite, soma or axon of another neuron; neurotransmitter is released from the presynaptic neuron to stimulate electrical excitation or inhibition in the postsynaptic cell

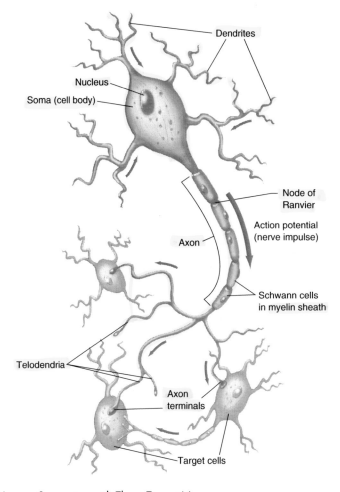

FIGURE 13.1 ◆ Motor Neuron Synapsing with Three Target Neurons

◆ **Exercise 13.1** *Identify Parts of the Neuron*

On a model of a neuron, identify the structures listed in Table 13.1 (Photo 367).

◆◆◆

II. Types of Motor Neurons

◆ **Exercise 13.2** *Examine Three Types of Motor Neurons*

1. **Giant Multipolar neurons** are located within the spinal cord. They send motor commands to voluntary skeletal muscles. View giant multipolar neurons under the microscope and, using Photos 366a) as a reference, identify the following structures: *soma, nucleus, dendrites, axon, Nissl bodies.* Sketch what you see in the space below.

2. **Pyramidal cells** are located within the cerebral cortex. These "upper" motor neurons send motor information from the brain to the cell bodies of motor neurons in the spinal cord. Pyramidal cells are named for the characteristic shape of their cell bodies. View a microscope slide of pyramidal cells and, using photo 366b as a reference, identify the following structures: *soma, nucleus, dendrites, axon.* Sketch what you see in the space below.

3. **Purkinje cells** are located in the cerebellum, which is involved in subconscious coordination of motor patterns. Using Photo 366c as a reference, identify the following structures: *soma, nucleus, dendrites, axon*. Sketch what you see in the space below.

Name _____

Lab 13: Nervous Tissue

◆ Practice

1. Answer the following questions about nervous tissue.

    a. What are the general characteristics of nervous tissue?

    b. What functions are performed by nervous tissue?

    c. How are the functions of nervous tissue reflected in its structure?

2. Describe the functions of the following parts of a neuron:

    a. dendrite

    b. soma

c. axon

d. axon hillock

e. myelin sheath

3. Draw and label the parts of a typical multipolar motor neuron in the space below.

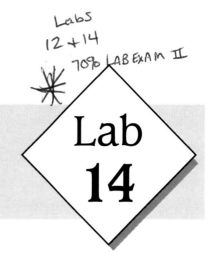

CENTRAL NERVOUS SYSTEM

Lab 14

Objectives

1. Identify major regions and specific structures on a human brain, sheep brain, model and diagram
2. Explain the functions of specific brain structures
3. Identify and describe the meningeal membranes
4. Label a diagram of whole spinal cord and a cross section through the spinal cord
5. Identify structures in a microscope slide of a spinal cord cross section

Materials

- Brain and spinal cord models
- Human brain specimens
- Sheep brain specimens
- Dissecting trays
- Disposable gloves
- Dissecting Instruments
- Microscope slide of spinal cord cross section

INTRODUCTION

The nervous system, like other body systems, can be divided into functional or structural classifications. For the purposes of laboratory study we will use the structural classifications of the nervous system, which includes the central nervous system (CNS) and peripheral nervous system (PNS). The CNS consists of the brain and the spinal cord. These structures contain groups of neuronal cell bodies, called **nuclei,** and bundles of neuronal processes, called **tracts.**

Functions of the CNS include initiation of motor commands and interpretation of sensory signals. Voluntary motor commands, which are sent to skeletal muscles, are initiated by neurons within the motor cortex of the cerebrum. Autonomic motor commands, which are sent to smooth muscle and glandular tissues, begin at the basal nuclei, diencephalon, brainstem and spinal cord. All motor commands are delivered to structures outside the CNS.

Sensory signals, on the other hand, are detected first by neurons whose cell bodies are located in the PNS. These first order sensory neurons deliver sensory information to second order sensory neurons located within the CNS. Incoming signals are then sent to cortical sensory areas for conscious perception and to various other CNS regions for determination of reflex activity.

CONCEPTS

I. The Brain

◆ **Exercise 14.1** *Examine a Whole Human Brain*

Cerebrum

View a whole human brain and a human brain model (Photo 368a, Figure 14.1). The brain is composed of several structures. The **cerebral hemispheres** comprise the main bulk of the brain, and house the functions of conscious thought, initiation of voluntary movement, somatic sensation and visceral sensation. The cerebrum is also instrumental in the special senses of vision, olfaction, and taste. It receives electrical signals from sensory receptors in your eyes, nose and tongue, and can identify the source of stimuli, such as a face, a rose or an orange.

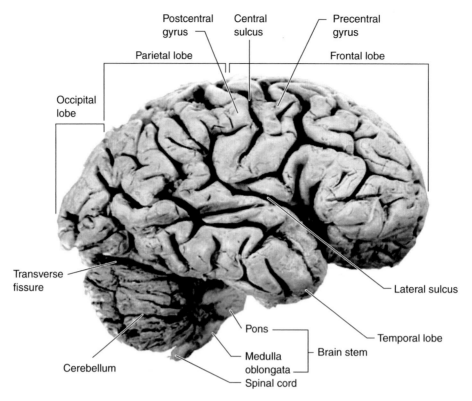

FIGURE 14.1 ◆ Lateral Brain

The cerebrum is covered with ridges, called **gyri** (singular is **gyrus**), and grooves, called **sulci** (singular is **sulcus**). This pattern allows for increased nervous tissue in a relatively small space. The gyrus immediately anterior to the **central sulcus** is called the **precentral gyrus,** and houses the primary motor cortex. Posterior to the central sulcus is the **postcentral gyrus,** which contains the somatosensory cortex. The surface of the cerebrum is composed of a few layers of neuronal cell bodies and their dendrites. This layer is called the **cerebral cortex,** and is responsible for the functions of the cerebrum. In the brain, regions of cell bodies and unmyelinated processes are called **gray matter.**

The cerebrum is divided into **lobes,** which are defined by certain landmarks. The **frontal lobe** extends from the anterior aspect of the brain to the central sulcus. The frontal lobe is the site of emotional affect, intelligence, cognition and voluntary movement. Motor speech is controlled from Broca's area within the frontal lobe. Posterior to the central sulcus and deep to the parietal bone is the **parietal lobe,** which is responsible for somatic sensation, gustatory sensation (taste), and associations related to these sensations. On one hemisphere, usually the left, is a region of the parietal lobe known as Wernicke's area that is responsible for understanding spoken and written language. The posterior most region of the cerebrum is the **occipital lobe,** and is responsible for vision. The **temporal lobe** is located ventral to the **lateral sulcus** and deep to the temporal bone. Olfaction is housed in this lobe, as well as structures important in memory.

View the dorsal surface of the cerebrum. The deep crevice between the hemispheres is termed the **longitudinal fissure** (also see Photo 370b). Deep within the longitudinal fissure is a thick axonal tract called the **corpus callosum** that connects the two hemispheres together. The corpus callosum allows coordination between the right and left sides of the cerebrum.

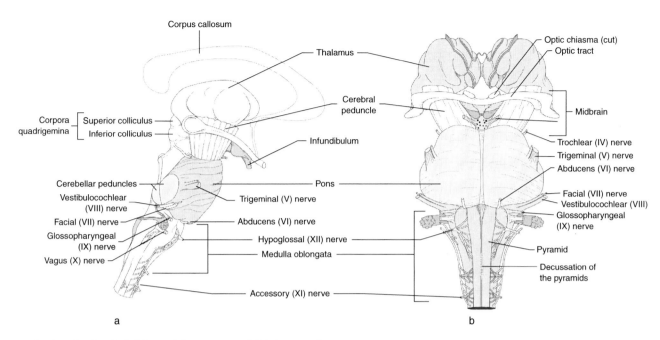

Corpus callosum

Thalamus

Optic chiasma (cut)
Optic tract

Cerebral
peduncle

Midbrain

Corpora
quadrigemina
Superior colliculus
Inferior colliculus

Infundibulum

Trochlear (IV) nerve
Trigeminal (V) nerve
Abducens (VI) nerve

Cerebellar peduncles
Vestibulocochlear
(VIII) nerve
Facial (VII) nerve
Glossopharyngeal
(IX) nerve
Vagus (X) nerve

Pons

Trigeminal (V) nerve

Abducens (VI) nerve

Hypoglossal (XII) nerve

Medulla oblongata

Facial (VII) nerve
Vestibulocochlear (VIII)
Glossopharyngeal
(IX) nerve

Pyramid

Decussation of
the pyramids

Accessory (XI) nerve

a

b

FIGURE 14.2 ◆ (a) Lateral Brainstem and (b) Anterior Brainstem

Cerebellum (Figure 14.1)

Located at the posterior and ventral aspect of the cerebrum is the bi-lobed and ridged **cerebellum.** This portion of the brain functions in coordinating and monitoring motor patterns. Like the cerebrum, the cerebellum has an outer surface composed of gray matter. The deep crevice between the cerebellum and cerebrum is the **transverse fissure.**

Brain Stem (Photos 369 and 370a, Figure 14.2)

The **Brain Stem** connects the cerebrum and cerebellum with the spinal cord. Nuclei and axon tracts are positioned within the three different regions of the brain stem.

The **midbrain** is located at the junction with the cerebrum and cerebellum. Posteriorly on the midbrain, and deep to the cerebellum, a set of four "bumps" define the **corpora quadrigemina.** The two **superior colliculi** of the corpora quadrigemina are involved in visual reflexes while the two **inferior colliculi** mediate auditory reflexes. Nestled superior and medial to the superior colliculi is the unpaired **pineal gland,** which secretes melatonin to help regulate sleep-wake cycles. Connecting the midbrain to the thalamus in the cerebrum are axonal tracts called **cerebral peduncles,** while the **cerebellar peduncles** are axonal tracts that connect the midbrain to the cerebellum.

Another region of the brainstem, the potbelly shaped **pons,** is positioned anteriorly on the brain stem. Extending inferiorly from the pons is the third region, or **medulla oblongata,** which is continuous inferiorly with the spinal cord. On the anterior surface of the medulla oblongata are two longitudinal ridges called the **pyramids.** The pyramids contain the **corticospinal tract,** also known as the pyramidal tract. This axonal tract carries voluntary motor commands from the motor cortex to lower motor neurons in the spinal cord. The distinction between the two pyramids becomes blurred at the point on the medulla oblongata known as the **decussation of the pyramids,** where axons cross from one side to the other.

Ventral Brain (Photo 368b, Figure 14.3)

A ventral view of the whole brain shows the ventral frontal and temporal lobes, and the anterior brain stem. Twelve pairs of **cranial nerves** are visible on this aspect. The **olfactory bulbs** receive innervation from **olfactory nerves (I)** originating in the nasal epithelium. Olfactory tracts extend from the bulbs posteriorly toward the hypothalamus. At the hypothalamic region is located the x-shaped **optic chiasma (II).** Extending distally from the optic chiasma are the **optic nerves,** while the **optic tracts** continue to the cerebrum via the thalamus. The remaining cranial nerves originate from the brainstem.

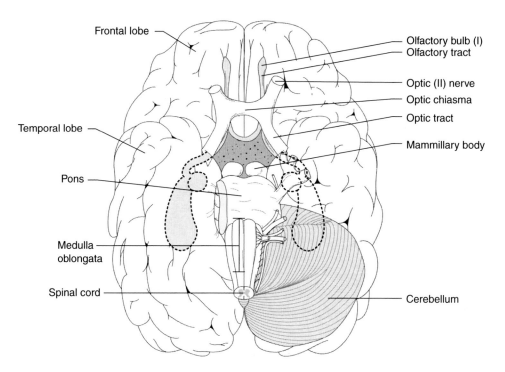

FIGURE 14.3 ◆ Ventral Brain

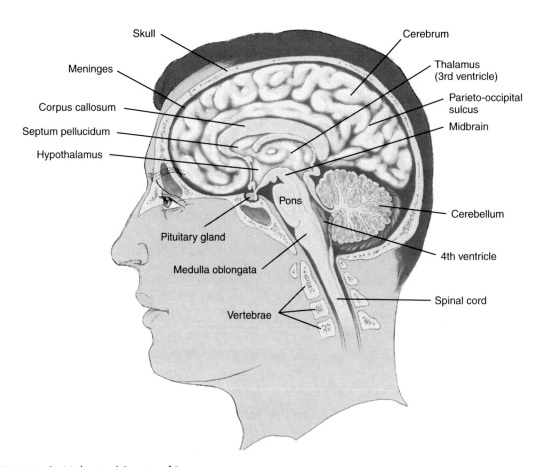

FIGURE 14.4 ◆ Midsagittal Section of Brain

Medial Brain and Ventricles (Photo 370a, Figure 14.4)

Observe a sagittal section of a human brain and a human brain model. The corpus callosum is cut in this view. A deep **parieto-occipital sulcus** separates the parietal and occipital lobes medially. Ventral to the corpus callosum on each hemisphere is the **lateral ventricle.** A thin membrane, the **septum pellucidum,** separates the two lateral ventricles in life. The **interventricular foramina** connect the lateral ventricles to the central, unpaired **third ventricle.** In a sagittal section the third ventricle appears as a shallow depression ventral to the lateral ventricle. The wall of the third ventricle is the medial aspect of the **thalamus.** This egg-shaped mass of gray matter integrates and directs ascending information to appropriate regions of the cerebrum for interpretation. A bridge of nervous tissue connects the two thalamic structures through the center of the third ventricle. The floor of the third ventricle is formed by the **hypothalamus.** Extending from the hypothalamus is a short stalk, the **infundibulum,** that suspends the **pituitary gland** within the sella turcica of the sphenoid bone. Extending from the third ventricle through the center of the midbrain is the duct-like **cerebral aqueduct** that terminates in the **fourth ventricle.** Continuing through the medulla oblongata and the spinal cord is a thin **central canal.**

Suspended from the roofs of each ventricle is a dark, granular-appearing mass that comprises the **choroid plexus.** This vascular structure filters fluid from blood plasma into the ventricles to become **cerebrospinal fluid (CSF).** CSF circulates between the ventricles and then leaves the fourth ventricle to bathe the outside of the brain.

Observation of nervous tissue color in a midsection of the cerebrum shows a distinction between the gray matter (unmyelinated structures) and the deeper **white matter** (myelinated fiber tracts) (Photo 370b). A medial view of the cerebellum shows thin lines of white matter within the gray matter that resembles a tree (the arbor vitae).

Meninges

The meninges are protective coverings of the brain (Photos 371a–371c). The innermost layer, the **pia mater,** is closely adhered to the gyri and sulci of the cerebrum. Embedded within the pia mater and tracing the course of the sulci are blood vessels. Superficial to the pia mater is a space that contains cerebrospinal fluid and filamentous, cobwebby strands that extend from the second meningeal layer, the **arachnoid mater.** The CSF-containing space between the arachnoid and pia mater is the **subarachnoid space.**

The most superficial meningeal membrane is the tough, double-layered **dura mater. Dural sinuses** are formed in locations where the two layers of the dura separate. Blood leaving the brain enters the dural sinuses and eventually drains into the jugular veins on the way to the heart. Small **arachnoid villi** push up into the dural sinus from the arachnoid layer, and allow CSF to return to the blood stream. A fold of dura mater called the falx cerebri fits deep down into the longitudinal fissure. Another fold, the tentorium cerebelli, separates the cerebrum from the cerebellum in the transverse fissure. If available, observe the dura mater in place on a human brain and removed from the brain.

◆◆◆

◆ Exercise 14.2 *Dissect a Sheep Brain*

Since the basic plan of all mammalian brains is the same, we can expand our understanding of the structure of the human brain by studying the brain of some other mammals. Because of its size and relative availability, the sheep brain is often used for comparative dissection. The sheep brain is somewhat smaller than the human brain but the structures are essentially analogous and the sheep brain is quite easy to work with. In this lab exercise, we will dissect the sheep brain and examine those structures common to both the sheep and the human brain. Use Photos 372a–373b and Figure 14.5 for reference.

1. Identify the various areas of the sheep brain (cerebrum, brain stem, diencephalon, cerebellum) and compare these areas to those of the human brain. Relatively speaking, which of these structures is obviously much larger in the human brain?

2. Place the sheep brain ventral surface down on the dissecting pan and observe the sac-like outmost layer of meninges—the **dura mater.** Feel its consistency and note its toughness. It is possible that the dura mater has been removed. If it is still present, gently remove it now. Be especially careful when separating the dura mater from the nerve tracts on the ventral surface of the brain. Notice that, like the human brain, its surface has fissures and gyri. Locate the middle layer of the meninges—the **arachnoid mater,** which appears on the brain surface as a delicate "cottony" material spanning the fissures (you may not be able to locate the arachnoid mater since this layer does not withstand preservation well). You should be able to locate the innermost layer of the meninges—the **pia mater,** which tightly adheres to the sulci and gyri of the brain.

3. Turn the brain so that its ventral surface is uppermost and identify the following structures:
   - olfactory tract
   - olfactory bulb (I)
     ▶ How does the size of these olfactory bulbs compare with that in humans? Explain why.

   - optic nerve (II)
   - optic chiasma
   - optic tract
   - midbrain
   - pons
   - medulla oblongata

4. Place the brain ventral side down on the dissecting pan and identify the following structures:
   - Gently force the cerebral hemispheres apart laterally to locate the **corpus callosum** deep in the longitudinal fissure
   - Carefully examine the **cerebellum.**
   - To expose the dorsal surface of the midbrain, gently spread the cerebrum and cerebellum apart. Identify the **pineal body.** Also identify the **superior** and **inferior colliculi,** collectively called the **corpora quadrigemina.**

5. The internal structure of the brain can only be examined after further dissection. Place the brain ventral side down on the dissecting pan and make a cut completely through it in a superior to inferior direction. Cut through the longitudinal fissure, corpus callosum, and midline of the cerebellum. Identify the following structures:
   - corpus callosum
   - septum pellucidum (pierce this membrane and probe the lateral ventricles)
   - lateral ventricle
   - choroid plexus (located inside the ventricles)
   - 3rd ventricle
   - cerebral aqueduct
   - 4th ventricle
   - thalamus
   - hypothalamus
   - pineal body
   - optic chiasma
   - pons
   - medulla

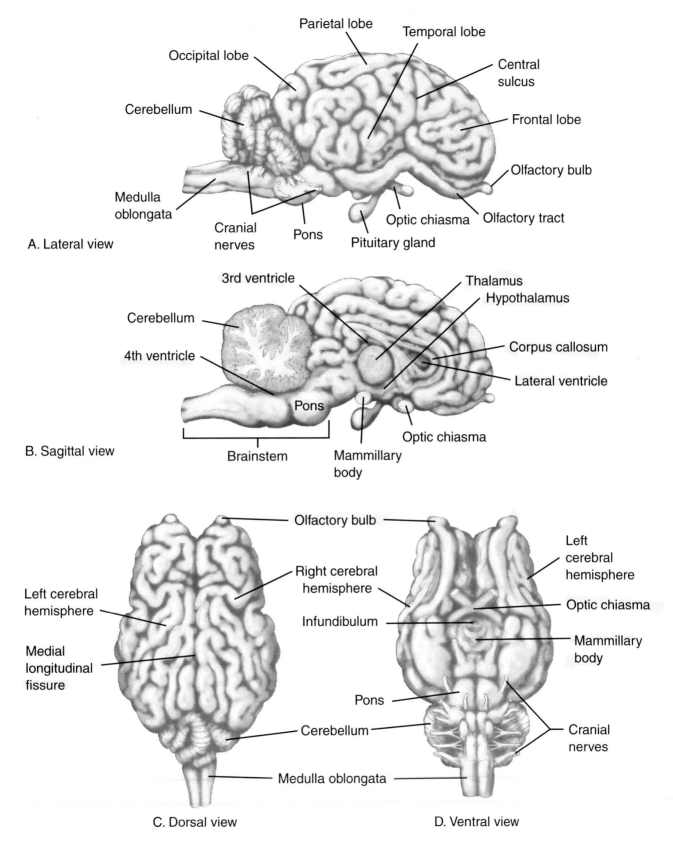

FIGURE 14.5 ◆ Sheep Brain

II. The Spinal Cord

The spinal cord is a highway for electrically coded information traveling from the brain to the body and from the body to the brain. Meningeal membranes of the brain continue inferiorly to surround the spinal cord, however the dura mater has only one layer in the spinal cord region. Cerebrospinal fluid circulates externally in the subarachnoid space and within the central canal at the center of the spinal cord. The meninges and the spinal cord are anchored and protected within the vertebral canal of the spinal column.

Although the spinal cord ends at the upper lumbar vertebrae, the meninges continue inferiorly to the lower lumbar region. The end of the spinal cord is called the **conus medullaris.** Extending from the conus medullaris is a connective tissue tether, the **filum terminale,** that anchors the spinal cord inferiorly to the coccyx or inferior sacrum.

Entering and exiting the spinal cord highway are **spinal nerves** that contain fibers from sensory and motor neurons. Sensory neurons (afferent neurons) deliver information about sensory stimuli from regions of the body to the spinal cord. Motor neurons (efferent neurons) carry motor commands from the spinal cord to the effectors, or muscles and glands, in the body. Spinal nerves exit the vertebral canal through intervertebral foramina. The spinal cord is shorter than the vertebral canal, thus the spinal nerves exiting from the sacral regions of the cord must travel inferiorly, beyond the conus medullaris, to exit through the foramina within the sacrum. Thus, below the level of the conus medullaris are several spinal nerves loosely contained within the meningeal envelope. These spinal nerves have the appearance of a horse's tail, and so are collectively called the **cauda equina.**

Because the upper and lower extremities of the body contain many independently controlled muscles, more integrating circuits are needed within the spinal cord at the regions that correspond to the arms and legs. This increase in nervous tissue is reflected in **cervical** and **lumbar enlargements** of the spinal cord.

Histologically, the spinal cord is a column of neuronal cell bodies surrounded by vertically oriented axonal tracts. Unlike the brain, the white matter within the spinal cord is superficial to the gray matter. Some axonal tracts bring electrical information from brain neurons to cell bodies within the spinal cord. These tracts are termed **descending tracts. Ascending tracts** carry information from the spinal cord to the brain. Tracts are named for their point of origin and end target. For example, the corticospinal tract delivers information from the motor cortex to the spinal cord, while the spinocerebellar tract transmits information from the spinal cord to the cerebellum. The bundle of axonal tracts dorsal, lateral and ventral to the gray matter are called the **dorsal (posterior) funiculus, lateral funiculus** and **ventral (anterior) funiculus** respectively.

In cross section, the gray matter is shaped roughly like a butterfly. The dorsal "wings" are called the **dorsal (posterior) horns,** and the ventral "wings" are the **ventral (anterior) horns.** Dorsal horns of gray matter contain cell bod-

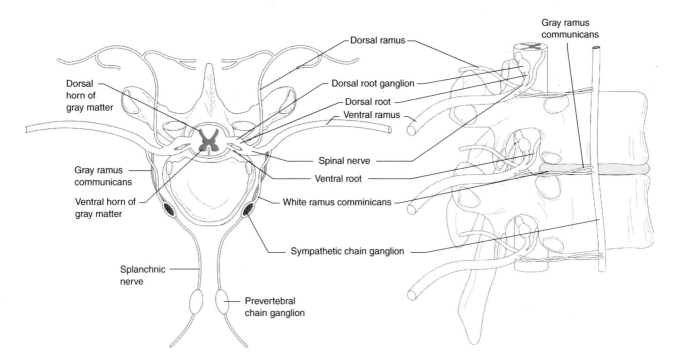

FIGURE 14.6 ◆ Cross Section of Spinal Cord in Spinal Column

ies of sensory neurons, while ventral horns are composed of motor neuron cell bodies. The **thoracolumbar region** of the spinal cord, from T$_1$ to L$_2$, contains a small **lateral horn** of gray matter composed of cell bodies related to the autonomic nervous system.

A **dorsal root** containing sensory neuron fibers joins the posterolateral surface of the spinal cord. The **ventral root** is composed of motor neuron fibers exiting the anterolateral spinal cord. The dorsal and ventral roots join together to form the spinal nerve prior to exiting through the intervertebral foramina.

First order sensory neurons are those that receive information from body sensory receptors. Cell bodies of first order sensory neurons are grouped in the dorsal root ganglia within each dorsal root. The cell bodies within the dorsal horn of the spinal cord are **second order sensory neurons.** First order axons may synapse on spinal cord second order neurons, or may ascend to synapse in nuclei within the brainstem, thalamus or cerebellum. Ascending tracts may carry sensory information from the same side of the body (**ipsilateral**) or from the opposite side of the body (**contralateral**). In other words, some ascending tracts on the left side of the spinal cord may carry sensory information from the left side of the body (ipsilateral), while other tracts on the left side of the cord may carry sensory information from the right side of the body (contralateral).

Upper motor neurons are neurons whose cell bodies are located in the cortex or in nuclei within the cerebrum or brainstem. Upper motor neuron axons descend within the spinal cord to synapse on cell bodies of **lower motor neurons** within the ventral horn. Axons from lower motor neurons exit the spinal cord through the ventral roots to synapse on effectors in the body.

Interneurons, or **association neurons,** are neurons within the central nervous system that make connections between two or more other neurons. Interneurons may carry sensory information to the contralateral side of the spinal cord.

◆ **Exercise 14.3** *Identify Structures in Spinal Cord Models*

Use Photo 374a as a reference and identify the following structures in a model of a cross section of spinal cord. For each region, record the type and parts of neurons present. The first has been completed for you.

Dorsal horn of gray matter

Cell bodies and dendrites of second order sensory neurons

(Receive) sensory information

Ventral horn of gray matter

(Sending) motor information

Lateral horn of gray matter

(Sending) sympathetic signals

Dorsal funiculus of white matter

Axons / Nerve fibers for communication
Contains posterior median sulcus

Lateral funiculus of white matter

Dorsal → Ventral roots come together along
lateral funiculus

Ventral funiculus of white matter

Contains Anterior median fissure

Dorsal root

Ventral root

Spinal nerve

Central canal

On a model of the spinal cord in place within the vertebral column observe the spinal nerves exiting the spinal cord through the intervertebral foramina.

◆◆◆

◆ **Exercise 14.4** *Identify Structures in a Microscope Slide of Spinal Cord Cross Section*

Examine a microscope slide of mammalian spinal cord cross section under low power. Sketch what you see in the space below, and label the features listed in Exercise 14.3.

Name _____

Lab 14: Central Nervous System

◆ Practice

I. Brain

    1. What structure is responsible for the production of cerebrospinal fluid into the ventricles?

    2. Trace the movement of CSF between ventricles. In what compartment outside the brain is CSF contained?

    3. How does CSF return to the venous blood?

    4. Write the functions for the following brain structures:

        a. superior colliculi

        b. inferior colliculi

        c. pineal gland

        d. corpus callosum

        e. decussation of the pyramids

        f. cerebral cortex

5. Meningitis is inflammation of the meninges due to a bacterial or viral infection. Why is this a serious condition? Given the close association of the meninges with the cerebral cortex, what symptoms might you expect for meningitis?

II. Spinal Cord

1. In the space below draw a cross section through the spinal cord and label the following structures: dorsal, lateral and ventral horns; dorsal, lateral and ventral funiculi, central canal, dorsal root, ventral root, spinal nerve.

2. In your own words, explain the reason for the presence of the cervical and lumbar enlargements.

3. What is the function of the filum terminale?

4. In what region of the spinal cord is a lateral horn present in the gray matter?

5. Match the following neuron parts with the appropriate location.

   A. First order sensory neuron cell bodies
   B. First order sensory neuron axons
   C. Second order sensory neuron cell bodies
   D. Second order sensory neuron axons
   E. Upper motor neuron cell bodies
   F. Lower motor neuron cell bodies
   G. Upper motor neuron axons
   H. Lower motor neuron axons
   I. Autonomic nervous system cell bodies

   _____ Dorsal horn

   _____ Ventral horn

   _____ Lateral horn

   _____ Dorsal root

   _____ Ventral root

   _____ Dorsal root ganglion

   _____ Ascending tracts in the funiculi

   _____ Descending tracts in the funiculi

   _____ Brain and brainstem

6. A spinal tap is a procedure in which a needle is inserted into the subarachnoid space of the spinal cord to withdraw CSF for diagnostic purposes. Why do you think it is important to insert the needle at the vertebral level of L_3 as opposed to higher levels?

PERIPHERAL NERVOUS SYSTEM

Lab
15

Objectives

1. Describe the structure of a nerve
2. Identify the connective tissue coverings in a nerve
3. Identify cranial nerves by number and name and state their functions
4. Identify the spinal nerves that contribute to the nerve plexuses
5. For each plexus, name terminal nerves and structures they innervate
6. Differentiate between root, spinal nerve, and ramus
7. Define **reflex,** and identify the parts of a reflex arc
8. Distinguish between somatic and autonomic reflexes

Materials

- Model and diagrams of a nerve
- Microscope slides of longitudinal and cross sectional nerve
- Model of spinal column with spinal nerves
- Reflex hammers
- Light pens

INTRODUCTION

The Peripheral Nervous System (PNS) includes sensory receptors and first order sensory neurons. In the PNS, groups of neuronal cell bodies are called **ganglia,** and bundles of axons are called **nerves.** Recall from Lab 14 that the cell bodies of first order sensory neurons are located in the **dorsal root ganglia (DRG),** found in the dorsal root of each spinal nerve. Other ganglia may contain cell bodies of autonomic motor neurons in addition to sensory neurons.

The function of the PNS is to detect sensory stimuli and convey that information to the CNS. Remember also that lower motor neurons from the spinal cord extend axons through peripheral nerves to deliver motor information to effectors.

CONCEPTS

I. Nerves

A nerve is a group of axons bundled together in the peripheral nervous system. A nerve is very similar to telephone cable running from a neighborhood. Each house has a telephone that must connect to a main switching station some distance away. It would be inefficient for the telephone wires from each house to take an independent path to the station. Instead, all of the wires from every house are packaged together in one large cable, which is then routed to the appropriate location. A nerve bundles together the axons from many cell bodies to efficiently route them to their end connections, or synapses.

The internal structure of a nerve consists of layers of connective tissue wrappings much like what is found in skeletal muscle. The individual axons are covered by an **endoneurium.** A bundle of axons, or **fascicle,** is surrounded by a **perineurium.** The fascicles, along with blood and lymph supply, are bound together by the **epineurium** to form the entire nerve.

Axons within peripheral nerves may be myelinated by **Schwann cells.** Cytoplasm is squeezed out of the Schwann cells as they wrap themselves around the axon. The cytoplasm becomes a lipid **myelin sheath** that electrically insulates the axon. Conduction velocity of action potentials is dramatically increased in myelinated axons. Bare spaces between Schwann cells along the axon are called **Nodes of Ranvier.** Action potentials jump from node to node in a form of saltatory conduction.

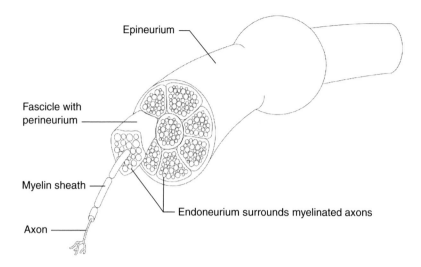

FIGURE 15.1 ◆ Peripheral Nerve

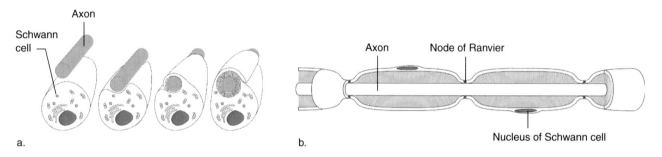

FIGURE 15.2 ◆ (a) Myelination of an Axon and (b) Myelin Sheath

◆ **Exercise 15.1**   *Examine Microscope Slides of Nerves*

Peripheral Nerve (longitudinal section) (Photo 374b)

   Identify the following structures and sketch what you see in the space below: *axon, Schwann cells, nodes of Ranvier*

1. Peripheral nerve (cross section) (Photos 375a and 375b)

Identify the following structures and sketch what you see in the space below: *axons, myelin sheaths, fascicles, endoneurium, perineurium, epineurium*

II. Cranial Nerves

The 12 pairs of cranial nerves (Table 15.1) primarily serve the head and neck. Cranial nerves I and II arise from the cerebrum, while the remaining nerves originate in the brain stem. The Vagus nerves (X) are the only cranial nerves that innervate structures in the thoracic and abdominal cavities. Disregarding autonomic fibers, cranial nerves may contain sensory fibers, motor fibers or both types.

Table 15.1
*Cranial Nerves**

◆ ◆ ◆

Number	Name	Sensory, Motor, or Both	Target Organ	Effect
I	Olfactory	Sensory	Chemoreceptors in nasal epithelium	Detect sense of smell
II	Optic	Sensory	Photoreceptors in retina of eye	Detect light and visual patterns
III	Oculomotor	Motor	Medial, superior and inferior rectus muscles; inferior oblique muscle of eyeball	Move the eyeball
IV	Trochlear	Motor	Superior oblique muscle of eyeball	Move the eyeball
V	Trigeminal	Both	Three divisions are present: V1: skin from tip of nose to top of head V2: skin of upper jaw and temporal region V3: skin of lower jaw and lip, muscles of mastication	Provide sensation to face and top of head, movement of chewing muscles
VI	Abducens	Motor	Lateral rectus muscle of eyeball	Move eyeball
VII	Facial	Both	Tongue and muscles of facial expression	Provide sensation of taste to anterior two thirds of tongue and moves muscles of facial expression
VIII	Vestibulocochlear	Sensory	Equilibrium and hearing receptors	Detect movement of head and sound
IX	Glossopharyngeal	Both	Tongue and swallowing muscles	Provide sensation of taste to posterior third of tongue and moves swallowing muscles
X	Vagus	Both	Thoracic and abdominal viscera, swallowing muscles	Provide parasympathetic innervation to viscera and moves swallowing muscles
XI	Accessory	Motor	Trapezius and sternocleidomastoid muscles	Movement of trapezius and sternocleidomastoid
XII	Hypoglossal	Motor	Intrinsic and extrinsic muscles of tongue	Movement of tongue

*Figures 14.2 and 14.3

III. Spinal Nerves

Each spinal nerve is formed by the convergence of a dorsal and ventral root extending from the spinal cord. The spinal nerve then branches into a **dorsal ramus** and **ventral ramus** (plural is **rami**). The dorsal rami serve structures in the back, and are typically short. The ventral rami serve the many structures anterior to the spinal column. In the cervical, lumbar and sacral regions of the spinal cord the ventral rami form many interconnecting branches that create a web-

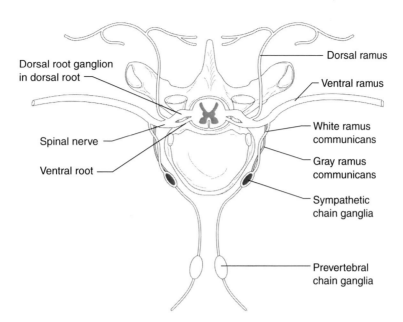

FIGURE 15.3 ◆ Rami of Spinal Nerves

like **plexus.** From the plexuses extend different terminal nerves that serve skin and muscle. Thus several spinal nerves may contribute to a specific terminal nerve. This design provides redundancy in the innervation of many structures, so that if one spinal nerve is damaged, structures will retain function from the remaining spinal nerves.

Also exiting the spinal nerve is a **white ramus communicans,** that provides a route for sympathetic axons to enter a series of stacked autonomic ganglia called the **sympathetic chain ganglia.** There is one sympathetic chain on each side of the vertebral bodies. Axons exit the chain ganglia via the gray ramus communicans. Other sympathetic fibers leave the spinal nerve and travel to an unpaired series of ganglia anterior to the vertebral bodies, called the **prevertebral chain ganglia.** Fibers arriving at the prevertebral chain are called **splanchnic nerves.**

There are 31 pairs of spinal nerves. Each spinal nerve is named in accordance to the level and region of the spinal cord from which it emerged. These nerves convey information between the CNS and all parts of the body (except the head and some areas of the neck) via receptors, muscles, and glands.

cervical spinal nerves C_1–C_8
thoracic spinal nerves T_1–T_{12}
lumbar spinal nerves L_1–L_5
sacral spinal nerves S_1–S_5
coccygeal spinal nerves C_o

Table 15.2
Nerve Plexuses and Terminal Nerves

[handwritten annotations: "Lab Practical — Test 4", "?'s from this chart"]

Plexus	Contributing Spinal Nerves • Specific Terminal Nerves	General Functions
Cervical	C_1–C_4	Sensory to back of head, front neck, upper shoulders; motor to neck
	• Phrenic Nerve	Motor supply to diaphragm muscle causes inspiration of air
Brachial	C_5–T_1	Sensory and motor to upper limb
	• Axillary	Sensory to shoulder skin, motor to deltoid and teres minor muscles
	• Radial Nerve	Sensory and motor to posterior arm, forearm and hand
	• Musculocutaneous Nerve	Sensory to anterior arm, motor to arm flexors
	• Median Nerve	Sensory to lateral 2/3 of hand, motor to forearm flexors
	• Ulnar Nerve	Sensory to medial 1/3 of hand, motor to forearm flexors
Lumbar	L_2–L_4	Sensory to skin of lower abdomen, lower back, external genitalia and thigh; motor to abdomen and medial and lateral thigh
	• Femoral Nerve	Sensory to skin of medial thigh, leg and foot, motor to quadriceps, sartorius and pectineus muscles
Sacral	L_5–S_4	Sensory to anterior and posterior thigh, buttock and external genitalia; motor to buttock, posterior thigh, leg and foot muscles
	• Sciatic Nerve	Composed of the tibial and common peroneal (common fibular) nerves
	• Tibial Nerve	Sensory to skin of posterior leg and sole of foot, motor to posterior thigh, leg and foot
	• Common peroneal	Sensory to skin of anterior leg and dorsum of foot, motor to lateral and anterior leg muscles

◆ **Exercise 15.2** *Identify Spinal Nerves in a Model of Spinal Column*

Review the exit of spinal nerves through intervertebral foramina in a spinal column model. Examine the position of the intervertebral discs. A *herniated disc* is a displaced intervertebral disc. How might this affect spinal nerves passing through intervertebral foramina?

◆ Exercise 15.3 | *Draw Spinal Cord Cross Section*

Draw a cross section of a spinal cord with the dorsal and ventral roots, spinal nerve, dorsal and ventral rami, and white ramus communicans. Use color pencils to draw and label the parts of lower motor neurons, first and second order sensory neurons, and sympathetic autonomic neurons.

IV. Reflexes

Reflexes are predictable, involuntary, unlearned responses to stimuli. Although you may be aware of some reflexes as they occur, all happen without conscious input. This is fortunate because our bodies rely upon many reflexes at all times to maintain homeostasis. If you had to concentrate on postural, digestive, blood pressure, vision, and other reflexes you would accomplish nothing else!

A **reflex arc** includes all of the structures involved in a specific reflex. There are many different reflex arcs, but all are comprised of a sensory receptor, sensory neuron, integration center, motor neuron(s) and effector. **Autonomic reflexes** involve stimulation of smooth muscle and glands by the autonomic nervous system. Although **somatic reflexes** engage skeletal muscles as effectors, they are involuntary. Conscious input from the brain is able to override some reflexes, such as elimination. Practice using biofeedback can increase the control an individual can exert over certain reflexes.

Spinal reflexes are those in which the integration center exists in the spinal cord. Again, no brain input is required for these reflexes to occur. Thus, patients who have sustained severe spinal cord injuries may still exhibit spinal reflexes inferior to the level of the injury. Testing certain reflexes is one way to demonstrate the integrity of specific nerves, or spinal cord regions.

◆ Exercise 15.4 | *Demonstrate a Somatic Reflex*

A **stretch reflex** occurs when the muscle spindle, or muscle stretch receptor, within a muscle tendon is activated. This induces a protective contraction of the same muscle to resist excess stretch. At the same time, antagonistic muscles are inhibited from contracting (**reciprocal inhibition**). Stretch reflexes can be elicited by exciting muscle spindles with a gentle tap to a tendon.

Patellar Reflex

Working in pairs, have one subject sit on a chair or desk so that the legs dangle. Palpate the small depression inferior to the patella and medial to the patellar tendon. Use a reflex hammer to quickly tap the patellar ligament at the region of this depression.

◆ **Exercise 15.5** *Demonstrate an Autonomic Reflex*

Pupillary Reflex

Smooth muscles within the iris (colored region) of the eye contract to dilate or constrict the size of the pupil, and thus control the amount of light that enters the eyeball. In a dim room, use a light pen to observe the effects of light and dark on the size of the pupil.

◆◆◆

Name _____

Lab 15: Peripheral Nervous System

◆ Practice

1. Complete the table of cranial nerves, target organs, and functions.

Cranial Nerve (number)	Target Organ(s)	Function
Trochlear Nerve ( )		
	Equilibrium and Hearing Receptors	
		Move muscles of facial expression
Accessory Nerve ( )		
		Sensation to face and movement of chewing muscles
	Thoracic and abdominal viscera	
(III)		
Olfactory Nerve (I)		
		Vision
	Intrinsic and extrinsic muscles of the tongue	
Abducens ( )		
		Taste to posterior tongue and movement of swallowing muscles

2. There are only seven cervical vertebrae; why are there eight cervical nerves?

3. Using the following options, fill in the table below for each spinal nerve plexus.

Terminal nerves:

sciatic   musculocutaneous

phrenic   median

ulnar   common peroneal

femoral   tibial

radial

Target organs:

A. Skin on posterior leg   H. Motor to leg flexors and ankle plantar flexors

B. Skin on anterior and medial thigh   I. Motor to leg extensors

C. Skin on lateral hand   J. Motor to wrist flexors

D. Skin on posterolateral arm and forearm   K. Motor to arm flexors

E. Skin on medial third of hand   L. Motor to ankle dorsi flexors and toe extensors

F. Motor to diaphragm muscle   M. Divides into two nerves

G. Motor to arm and wrist extensors

Nerve Plexus	Nerves	Target
Cervical Plexus		
Brachial Plexus		
Lumbar Plexus		
Sacral Plexus		

4. Draw the components of the patellar stretch reflex. Indicate the specific name of each part of the reflex arc (e.g., name the specific muscle effector).

5. What is the primary distinction between somatic and autonomic reflexes? How are they similar?

ANATOMY OF EYES AND EARS

Lab
16

Objectives

1. Identify and explain the functions of the important extrinsic and intrinsic anatomical features of the eye in diagrams, models and a cow eyeball
2. Identify and explain the functions of the important anatomical features of the external, middle and inner ear on diagrams and models

Materials

- Cow eyeballs
- Dissecting trays
- Dissecting instruments
- Disposable gloves
- Human eye model and diagrams
- Human ear model and diagrams

INTRODUCTION

We know that sensory receptors are located throughout the body to provide the central nervous system with information about our internal and external environment. When someone says "senses," however, most of us think of the **special senses,** which include taste, smell, vision, hearing, and equilibrium. At the most elemental level, the sensory organs detect sensory stimuli and transmit electrical signals to the brain. Our sensory experience includes more than merely detecting a stimulus, but involves complex processes of identification, associations and memories.

In this lab, we will examine the anatomy of eyes and ears, which house receptors for the special senses of vision, hearing and equilibrium.

CONCEPTS

I. The Eye

Structures associated with vision can be classified into two groups: accessory structures and intrinsic structures. **Accessory structures** are extrinsic to the eyeball, and contribute indirectly to vision. **Intrinsic structures** are located within the eyeball itself.

Intrinsic Structures of the Eye

The eyeball is essentially composed of three layers, or **tunics.** The outermost tunic is the protective **fibrous tunic** that includes the sclera and cornea. Extrinsic eye muscles are anchored into the tough, white **sclera,** which shapes and protects all but the anterior most portion of the eye. Anteriorly the **cornea** forms a bulging, transparent window for light transmission into the eyeball.

The middle layer, the **vascular tunic,** is composed of the choroid, the ciliary body and the iris. Blood vessels within the **choroid** supply nutrients to all of the tunics within the eye. Melanin, a brown pigment, is present in the choroid to absorb light that might otherwise interfere with vision. The choroid is continuous with the **ciliary body,** which encircles the disk-shaped **lens.** Tiny ciliary muscles in the ciliary body control the lens shape by stretching or relaxing the attached **suspensory ligaments.** At rest, the ciliary muscles stretch the suspensory ligaments to flatten the lens and allow distant vision. When ciliary muscles contract, the suspensory ligaments relax and allow the lens to bulge, accommodating near vision. Anteriorly the vascular tunic becomes the **iris.** Melanin within the iris causes a person's eyes to appear a certain color. In the center of the iris is the **pupil,** which is an opening to allow the passage

Table 16.1
Accessory Structures of the Eye

◆◆◆

Structure	Function
Eyebrows	Direct sweat away from eye
Eyelids	Prevent substances from entering eye, spread tears across eye surface for moisture
Conjunctiva	Lines inside eyelids and anterior eyeball; produces lubricating mucus
Lacrimal apparatus	Secretes lacrimal fluid (tears) onto surface of eye for moisture; lysozyme inhibits bacterial growth, tears wash substances away from eye surface
Extrinsic eye muscles	
Superior rectus	Moves eye superiorly
Inferior rectus	Moves eye inferiorly
Lateral rectus	Moves eye laterally
Medial rectus	Moves eye medially
Superior oblique	Moves eye inferiorly and laterally
Inferior oblique	Moves eye superiorly and laterally

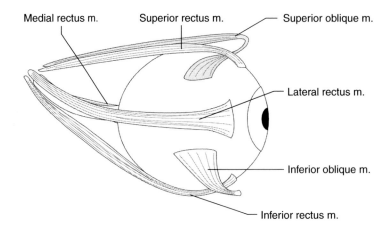

Medial rectus m. Superior rectus m. Superior oblique m. Lateral rectus m. Inferior oblique m. Inferior rectus m.

FIGURE 16.1 ◆ Lateral View of Extrinsic Eye Muscles

of light to the interior of the eyeball. Radial smooth muscles, like the spokes of a bicycle wheel, surround the pupil and, when stimulated by the sympathetic nervous system, dilate the pupil. Concentric circular smooth muscles within the iris are stimulated by the parasympathetic nervous system to constrict the pupil.

The double layered **retina** forms the inner **sensory tunic.** The superficial pigmented layer lines the choroid and covers the ciliary body and posterior iris. Internal to the pigmented layer is the neural layer, which contains **photoreceptors. Rods** are photoreceptors that detect even the smallest quantity of light, but that do not distinguish between different wavelengths. **Cone** photoreceptors require more light to be activated, and detect specific ranges of wavelengths, or colors. Images detected by cones are also more defined and distinct than when detected by rods. Rods are scattered throughout the neural layer to provide peripheral vision. Most cones are positioned in a tiny depression in the retina called the **fovea centralis,** which is centered in the yellow, oval **macula lutea** on the inner posterior surface. To see sharp images in color, light must be centered on the fovea centralis. Within the retina, photoreceptors pass sensory information to short bipolar cells, which then excite ganglion cells. Axons from ganglion cells travel along the inner surface of the retina and exit the eyeball at the posterolaterally located **optic disc.** The optic disc is the **blind spot,** because no photoreceptors are present at this location. The axons are bundled together to form the **optic nerve.**

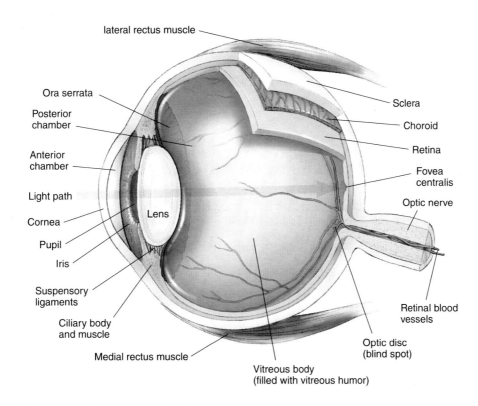

FIGURE 16.2 ◆ Transverse Section through the Eyeball

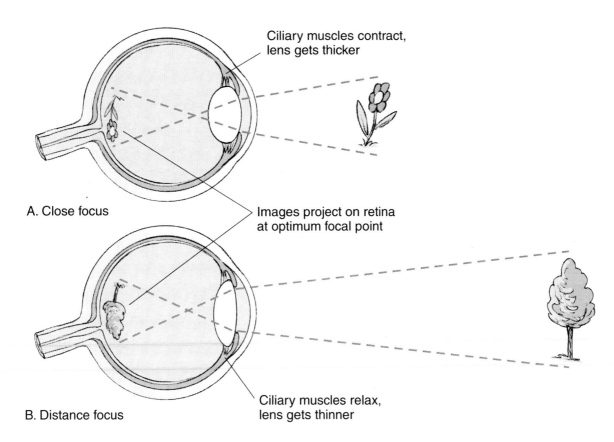

FIGURE 16.3 ◆ Focusing

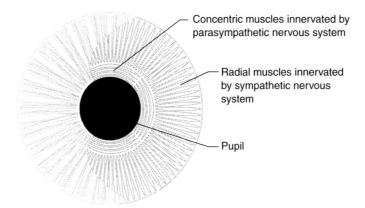

— Concentric muscles innervated by
parasympathetic nervous system

— Radial muscles innervated
by sympathetic nervous
system

— Pupil

FIGURE 16.4 ◆ Iris

◆ Exercise 16.1 *Find Your Blind Spot*

Although the blind spot in your eye creates deficits in your vision, you do not notice its effects because your brain fills in the gap. You can *see* the effect of your brain's supplementing activity by performing this exercise. Close or cover your right eye. With the left eye, focus on the round dot in Figure 16.5. Hold this page away from you and slowly move it toward your face while continuing to focus on the dot. When the gap in the line is focused on your blind spot, your brain "fills in" the gap based on surrounding visual cues, and the line appears unbroken in your peripheral vision.

FIGURE 16.5

◆ Exercise 16.2 *Examine Structures in a Cow Eye*

1. Obtain a preserved cow eye and your dissecting instruments.
2. Examine the external anatomy of the eye. Note specifically the sclera, cornea, and optic nerve. The eye may still retain a cushion of adipose tissue that protects the eye in the orbit of the skull.
3. Trim away any adipose tissue. Using scissors, make an incision slightly lateral to the cornea, and continue cutting around the cornea to remove it from the eyeball. (Photo 376a)
4. The black pigmented area surrounding the lens is the ciliary body. Can you detect the suspensory ligaments attaching the lens to the ciliary body?
5. The pupil is the hole in the center of the iris, which is a continuation of the ciliary body.
6. Carefully remove the lens from the eyeball. Note its consistency.
7. The vitreous humor within the posterior compartment is a viscous fluid that amplifies the image and helps the eyeball to retain its shape.
8. The retina is a shiny membrane on the inner surface of the posterior compartment. Find the macula lutea and the optic disc.
9. When you have completed your examination of the eyeball, discard it in the bag provided.

◆ **Exercise 16.3** *Examine a Model of the Human Eye*

Obtain a model of the human eye and observe the following structures:

- Extrinsic eye muscles
- Sclera
- Optic nerve
- Choroid
- Ciliary body
- Lens

- Iris
- Pupil
- Retina
- Optic disc
- Macula lutea with fovea centralis

II. The Ear

The ear is divided into three regions: the external, middle and inner ear. The **auricle** of the external ear acts like a radar dish to direct sound waves into the **external auditory canal.** The eardrum, or **tympanic membrane** is the junction between the external and middle ear.

Sound waves vibrate the thin tympanic membrane. Energy of the sound wave is converted into mechanical energy by the tympanic membrane, and is further amplified by the vibration of the three tiny **ossicles,** the smallest bones in the human body, within the middle ear. Vibrations of the tympanic membrane move the **malleus,** which vibrates the **incus,** which in turn is connected to the **stapes** (Photo 376b). Two small skeletal muscles, tensor tympani and stapedius, reflexively dampen the motion of the malleus and the stapes upon the onset of loud sounds. The **pharyngotympanic tube** exits the middle ear and opens into the pharynx (throat) region in the soft palate. This opening allows pressure to equalize within the middle ear. Unfortunately, bacteria and viruses can migrate up the pharyngotympanic tube to take up residence in the middle ear and cause an ear infection (otitis media).

The middle ear is bounded medially by the temporal bone. The inner ear is a labyrinthine structure embedded within the temporal bone. Two small windows in the bony wall between the middle and inner ear are covered by thin membranes. The stapedius vibrates against the **oval window.** The **round window** is located inferior to the oval window.

Medially, the inner ear is a beautiful and complex structure that houses receptors for both hearing and equilibrium. The **bony labyrinth** is lined by a **membranous labyrinth,** which is composed of two parts: the vestibular

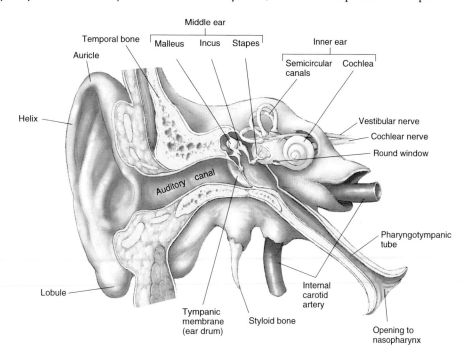

FIGURE 16.6 ◆ Structures of the Ear

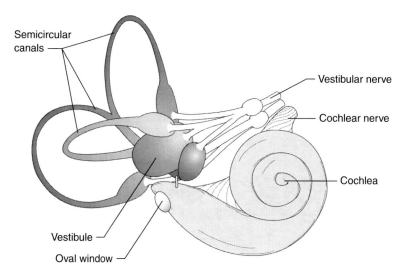

Semicircular canals

Vestibular nerve

Cochlear nerve

Cochlea

Vestibule

Oval window

FIGURE 16.7 ◆ Membranous Labyrinth

apparatus and the cochlea. Exiting from the vestibular apparatus, the **vestibular nerve** quickly joins with the **cochlear nerve** to form the **vestibulocochlear nerve** (VIII).

A sac-like region within the vestibular apparatus, the **vestibule,** contains two receptors that detect linear motion. These are the utricle and saccule. Extending from the vestibule are three tubes collectively called the **semicircular canals.** Fluid within the semicircular canals excites receptors called the cristae ampullares, which provide information about angular movements of the head.

The **cochlea** is a spiral system of ducts wound about a bony spiral wall, called the modiolus. A cut through the cochlea reveals three ducts; the **scala vestibuli** and **scala tympani** both contain **perilymph** fluid, while the centrally located **cochlear duct** contains **endolymph** fluid. Housed within the cochlear duct is the hearing receptor, called the **spiral organ of Corti.**

The spiral organ of Corti lies on the floor of the cochlear duct, and extends the length of the duct to the apex of the spiral. Supporting cells along the duct floor, or basilar membrane, surround **hair cells,** from which project cilia into a stiff **tectorial membrane.** Because the scala vestibuli ends at the oval window, vibrations from the stapes creates fluid waves in the endolymph that move along the scala vestibuli to the scala tympani. The scala tympani ends at the round window. Fluid waves bounce from the oval window to the round window much like the waves in a waterbed move from one side to the other. As the endolymph in the scala tympani moves up and down, the basilar membrane with its supporting cells and hair cells also moves. The cilia, however, are locked within the tectorial membrane and are forced to bend. Bending of the cilia excites each hair cell, and action potentials are generated within the sensory neurons associated with hair cells. In this way, the energy from the fluid wave is converted to electrical energy of an action potential.

◆ Exercise 16.4   *Examine a Model of the Human Ear*

Obtain a model of the human ear and identify the following structures:

- Auricle
- Helix
- Lobule
- External auditory canal
- Tympanic membrane
- Ossicles: incus, malleus, stapes
- Pharyngotympanic tube

- Round window
- Oval window
- Vestibule
- Semicircular canals
- Cochlea
- Vestibulocochlear nerve

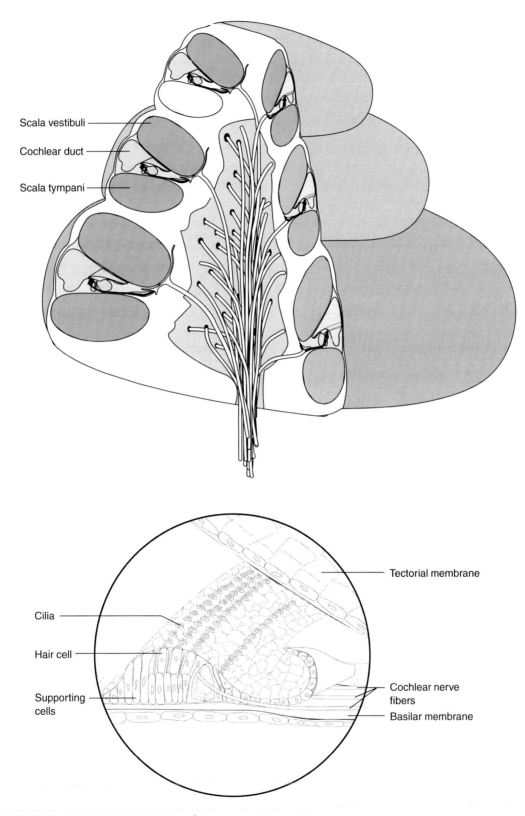

Scala vestibuli

Cochlear duct

Scala tympani

Tectorial membrane

Cilia

Hair cell

Supporting cells

Cochlear nerve fibers

Basilar membrane

FIGURE 16.8 ◆ Cochlea and Spiral Organ of Corti

Name _____

Lab 16: Anatomy of Eyes and Ears

◆ **Practice**

1. Label as many structures as you can in the diagrams of the eye and ear.

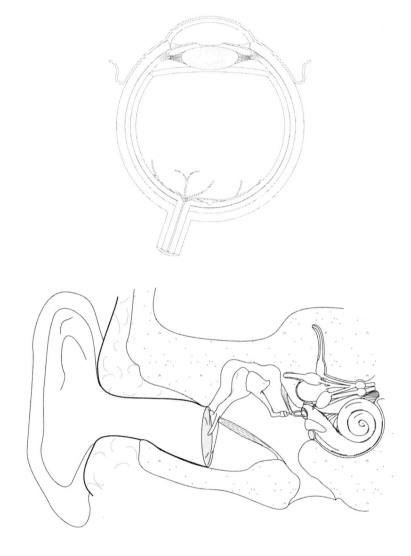

2. Why do you have a "blind spot" in your vision?

3. What is the function of the lens? Explain how accommodation functions to allow near vision.

4. Describe the placement of rods and cones in the retina. What is the function of the fovea centralis?

5. Which type of photoreceptor requires the least amount of light?

6. Which type of photoreceptor provides the clearest vision and color images?

7. List the functions of the following extrinsic eye structures:

   lateral rectus muscle

   superior oblique muscle

   inferior rectus muscle

conjunctiva

eyelid

lacrimal apparatus

8. What is the function of the ossicles?

9. Why is the middle ear prone to infection when a person has a sore throat?

10. List the different types and locations of equilibrium receptors.

ENDOCRINE GLANDS AND HORMONES

Lab
17

Objectives

1. Identify the major endocrine organs in human diagrams and torso models
2. Identify the major endocrine organs within a cat specimen
3. Understand the actions and target organs of specific hormones
4. List the hormones synthesized and secreted by each major endocrine organ

Materials

- Cat specimens
- Dissecting instruments
- Disposable gloves
- Torso model
- Video: *Selected Actions of Hormones and Other Chemical Messengers*, Dr. J. Barrena, Dr. R. Vines and University Media Services, CSU, Sacramento, Benjamin/Cummings Pub. Co., Inc. 1994.

INTRODUCTION

Endocrine glands, together with the hormones that they produce, comprise the endocrine system. This system works together with the nervous system to control homeostasis. Recall that homeostasis is maintained by many different control systems within the body, called feedback systems. Feedback systems can be positive or negative in nature. Positive feedback systems tend to amplify the stimulus while negative feedback systems will reverse the stimulus. Hormones are used within feedback systems as regulatory chemicals that act both as messengers to communicate physiological conditions, and as effectors to cause particular physiological events.

The regulation of blood glucose levels is a negative feedback system, as are most of the body's feedback systems. The hormone insulin is released from the pancreas in response to high blood glucose levels, for example after a person has eaten a meal. Insulin allows glucose to be stored or used by the body's cells. If blood glucose levels are low, insulin is not produced, but instead the hormone glucagon is secreted by the pancreas to release stored glucose from specific body cells into the blood. Thus, blood glucose levels are homeostatically maintained using two hormones.

Labor and delivery, on the other hand, is regulated by a hormone in a positive feedback system. Stretching of the pregnant uterus stimulates a release of the hormone oxytocin from the pituitary gland. Oxytocin causes contractions of the uterus, which in turn stimulate release of additional oxytocin and additional uterine contractions. The process continues and escalates until the baby is expelled from the uterus during delivery.

CONCEPTS

I. Gross Anatomy of Endocrine Glands

Of all the endocrine organs in the human body, the relationship between the **hypothalamus** and **pituitary** glands is the most complex. Recall that the hypothalamus is a neural structure within the floor of the third ventricle in the brain. Some of the neurons present produce hormones in the cell bodies, and store the hormones in the axon terminals. Additionally, epithelial secretory cells are present in the hypothalamus that secrete hormones into the blood.

Attached to the hypothalamus by a stalk-like infundibulum is the pituitary gland. This lobular gland sits in the sella turcica of the sphenoid bone in the skull. Secretions from the hypothalamus regulate functions of the pituitary gland. The pituitary gland is also called the **hypophysis**. The anterior lobe of the hypophysis is comprised of secretory cells, and is called the **adenohypophysis**. The posterior lobe contains the axon terminals from hypothalamic neurons, and is called the **neurohypophysis**.

209

Table 17.1
Hormones of the Hypothalamus and the Pituitary

◆◆◆

Hypothalamic Hormone	Effect	Pituitary Hormone	Target Organ and Effect
Released from Secretory Tissue:			
Human Growth Hormone Releasing Hormone (GHRH)	Stimulates adenohypophysis to secrete Growth Hormone	Growth Hormone (GH)	Stimulates somatic growth in bones and tissue
Human Growth Hormone Inhibiting Hormone (GHIH)	Inhibits the adenohypophysis from secreting Growth Hormone		
Thyrotropin Releasing Hormone (TRH)	Stimulates the adenohypophysis to secrete Thyroid Stimulating Hormone	Thyroid Stimulating Hormone (TSH)	Stimulates thyroid gland to secrete thyroxine and triiodothyronine (together called Thyroid Hormone)
Corticotropic Releasing Hormone (CRH)	Stimulates the adenohypophysis to secrete Adrenocorticotropic Hormone	Adrenocorticotropic Hormone (ACTH)	Stimulates the adrenal cortex to secrete glucocorticoids and testosterone
Gonadotropin Releasing Hormone (GnRH)	Stimulates the adenohypophysis to secrete Follicle Stimulating Hormone and Luteinizing Hormone	Follicle Stimulating Hormone (FSH), Luteinizing Hormone (LH)	LH—stimulates ovulation and secretion of estrogen and progesterone in ovaries; stimulates testosterone secretion from testes FSH—stimulates follicle and oocyte (egg) maturation in ovaries, stimulates sperm production in testes
Prolactin Releasing Hormone (PRH)	Stimulates the adenohypophysis to secrete Prolactin	Prolactin (PRL)	Stimulates lactation in breast secretory tissue
Prolactin Inhibiting Hormone (PIH)	Inhibits the adenohypophysis from secreting Prolactin		
Synthesized in Neuronal Cell Bodies:			
Oxytocin	Released from axon terminals in the neurohypophysis		Stimulates contraction of the uterus, initiates milk release from breast secretory tissue
Antidiuretic Hormone (ADH)	Released from axon terminals in the neurohypophysis		Stimulates kidney cells to reabsorb more water, making urine more concentrated

Table 17.2
Major Endocrine Organs and Hormones and Their Effects

◆◆◆

Endocrine Organ	Location	Major Hormones	Target Organ and Effects
Pineal Gland	Roof of the third ventricle in the brain	Melatonin	High levels stimulate brain neurons and cause drowsiness, low levels allow alertness
Thyroid Gland	Anterior surface of the trachea inferior to the larynx	Thyroxine (T_4) Triiodothyronine (T_3)	T_4 is converted to T_3. T_3 increases metabolism of most body cells
Parathyroid Glands	Three or four pairs on posterior surface of thyroid	Parathyroid Hormone (PTH)	Increase calcium absorption from small intestine, stimulate osteoclast activity to release calcium from bone, increase kidney reabsorption of calcium
Adrenal Gland	Sits on top of kidney • Cortex (outer region) • Medulla (inner region)	Cortex: • Aldosterone • Cortisol • Testosterone Medulla: • Epinephrine • Norepinephrine	 Stimulates kidneys to absorb more sodium Increase blood glucose, fatty acids and amino acids Stimulates puberty, male secondary sex characteristics Epinephrine and norepinephrine increase blood glucose, increase heart rate, increase blood pressure (fight or flight response)
Thymus	Deep to sternum	Thymopoietin Thymosin	Both cause immunocompetence in T-cells of immune system
Pancreas	Inferior and to the left of the stomach	Insulin Glucagon	Insulin transfers glucose from blood to body cells Glucagon releases stored glucose from glycogen in liver and skeletal muscle cells
Ovaries	Lateral to uterus in pelvic cavity of females	Estrogen Progesterone	Estrogen—development of female secondary sex characteristics Both—regulate menstrual cycle
Testes	Within scrotum of males	Testosterone	Development of secondary sex characteristics, maintenance of sex drive, development of sperm

Axons extending from hypothalamic neurons to the neurohypophysis run through the infundibulum as the **hypothalamic-hypophyseal tract.** Hormones synthesized in hypothalamic neuronal cell bodies travel down the tract to be stored in the axon terminals of the neurohypophysis.

Capillary beds surround both the hypothalamus and the pituitary gland. The **hypophyseal portal system** is a connection between the hypothalamic capillaries and the pituitary capillaries. Hormones from the hypothalamus enter the hypothalamic capillaries and are transported directly to the pituitary gland. At the pituitary capillaries, hypothalamic hormones can exit the blood stream to have an effect on the adenohypophysis. Hormones secreted by the pituitary can enter the pituitary capillary bed and will then be transported to the heart, which distributes the hormones systemically.

◆ **Exercise 17.1** *Identify the Major Endocrine Organs on a Human Torso Model*

Use a human torso model to review the position of the major endocrine organs listed in Tables 17.1 and 17.2. Review the function of each organ as you identify it.

◆◆◆

◆ **Exercise 17.2** *Dissect and Observe Endocrine Organs in a Cat Specimen*

1. Obtain a preserved cat from your instructor. Your dissection team will use this same cat throughout the semester so label it with the provided tags. Note that the cats have been injected with colored latex for ease in distinguishing arteries (red) and veins (blue).
2. Lay the cat on its back on a dissecting tray. Make a vertical incision starting below the chin and continuing on one side of the sternum to the inferior border of the sternum. From the top and bottom of the vertical incision, make lateral incisions to both the right and left sides.
3. Use scissors to cut the ribs through the vertical incision where they attach to the sternum. Carefully open up the chest cavity. You will have to break the ribs to fold them open and away from the midline.
4. Within the thoracic cavity, observe the centrally located heart within the pericardium. On both sides of the heart you will see the lobes of the lungs. Superior to the heart, and intermediate to the right and left lungs you will see the lobular **thymus** gland. (Photo 409)
5. Find the white, cartilage-ringed trachea deep to the thymus. At the superior aspect of the trachea you will find the glandular tissue of the **thyroid.** (Photo 411) This gland sits on top of the larynx, or voice box region.
6. Make a midline incision down the length of the abdomen to a point about 2 inches from the junction between the back legs. Make two angled incisions from the bottom of the first cut to the junctions between the abdomen and the back legs. **Do not cut too deeply or you will damage the underlying organs.**
7. Reflect back the muscle walls of the abdomen to expose the abdominopelvic organs. Use the Photos 409 and 410a to assist with initial identifications. Observe the dark, large liver just inferior to the diaphragm muscle. The spleen has the same coloring as the liver and can be seen extending from the left side of the cat. The intestines are covered by a membranous omentum hanging off the inferior edge of the stomach.
8. Carefully lift the omentum superiorly to expose the small intestines. **Do not tear or remove any organs or the omentum at this time!** The **pancreas** is a subtle tan organ deep to the omentum and inferior to the stomach (Photo 410b). It is very granular in appearance and to the touch.
9. Replace the omentum and then **gently** move the intestines to one side to expose a kidney lying against the posterior abdominal wall (Photo 412). The **adrenal** glands are tiny, granular organs encased in fat at the superior aspect and slightly medial to each kidney.
10. If your cat is a female you can find the **ovaries** at the ends of each V-shaped uterine horn (Photo 414). The uterus is located against the posterior abdominal wall, and extends in a V-shape from deep to the urinary bladder.

11. In the male cat the **testes** are located in the external scrotal sac (Photo 413). Carefully cut open one side of the scrotum to expose the small white testis within. You do not need to identify other reproductive structures at this time.

    Note: Make sure you examine both a male and female cat.

◆◆◆

II. Hormone Functions

Hormones function by changing the activity of a target cell or organ. Each hormone has a specific shape, and can only bind to a uniquely shaped receptor. Hormones will circulate in the blood and tissue throughout the body, but will only have an effect on target cells which contain the specific receptor to which it can bind. Effects depend upon the type and location of the target cells.

Because it is difficult and costly to maintain live animals for laboratory purposes, we will study the effects of some specific hormones by viewing a video and answering questions regarding the experiments observed.

◆ Exercise 17.3   *View Experiments Showing Hormone Actions*

The following experiments are demonstrated on the video, *Selected Actions of Hormones and Other Chemical Messengers.* Watch each experiment and answer the questions regarding hormone function.

A. Effect of Pituitary Hormones on Egg Production in the Frog

Experimental Protocol

1. Two female frogs are examined for evidence of ovulation by gently squeezing the abdomen toward the chloacal opening. Neither frog is currently ovulating.
2. The experimental frog is injected with a commercially prepared frog pituitary extract into the abdominal cavity.
3. The control frog is injected with the same volume of normal saline solution into the abdominal cavity.
4. Following the injections, the frogs are returned to jars containing a small amount of pond water.
5. 48 hours later the frogs are both again examined for evidence of the presence of eggs in the chloaca.

Questions

1. Which frog underwent ovulation?

   experimental frog c pituitary extract

2. What is the purpose of the control frog?

   to show effects of extract

3. What is the function of Luteinizing hormone in male and female humans?

4. What is the effect of Follicle Stimulating hormone in male and female humans?

5. Which hormone might be used to stimulate ovulation in a woman wishing to undergo *in vitro* fertilization?

6. What endocrine gland secretes both LH and FSH?

B. Effect of Hyperinsulinism

Experimental Protocol

1. A goldfish is placed in Bowl A, which contains an insulin solution (400 IU insulin /100 ml water). After 5 minutes the fish's body begins to twitch.
2. For several minutes the fish swims in circular movements but is unable to maintain an upright position.
3. After 20 minutes the fish is on its side in a comatose condition.
4. The fish is transferred to Bowl B, which contains a 10% solution of glucose.
5. After 5 minutes the fish begins to revive.
6. After 15 minutes the fish resumes its normal swimming activity.

Questions

1. What is the normal function of insulin?

   To convert sugar

2. Does the brain require insulin for uptake and use of glucose?

   yes

3. Why did the fish enter the coma while in Bowl A?

   Too much insulin promoting blood to use sugar.
   Brain is deprived of sugar.

4. What was the mechanism of revival when the fish was transferred to Bowl B?

   Sugar enters blood leveling out sugar levels

5. How does this experiment illustrate the dangers of insulin overdose in a diabetic?

   Can cause coma and even death

6. If a diabetic is without insulin for an extended period of time, he may enter a coma. How is this diabetic coma different than the insulin shock described in the experiment?

Coma is caused by hypoglycemia
Shock is caused by hyperglycemia

7. From what endocrine organ is insulin secreted?

Pancreas

C. Effect of Epinephrine and Acetylcholine on the Heart

Experimental Protocol

1. A frog is pithed to destroy the brain. A preparation exposes the heart as it is beating within the thoracic cavity of the frog. The heart is connected to a pen-writer to record the force of contraction and heart rate (HR).
2. A base-line HR is determined in beats per minute (bpm).
3. The heart is flushed with a 1:1000 solution of epinephrine.
4. After a few minutes the heart is then flushed with a 1:50 solution of acetylcholine (ACh).

Questions

1. What is the effect of epinephrine on HR and force of contraction of the heart beat?

↑ HR – much more forceful

2. What is the effect of acetylcholine on HR? (Note that acetylcholine is not classified as a hormone.)

↓ HR by affecting   *force of contraction decrease*

3. From what endocrine gland is epinephrine secreted? From which region of the gland?

Adrenal Gland

4. Epinephrine functions in which branch of the autonomic nervous system?

Sympathetic

5. ACh functions in which branch of the ANS?

Name _____

Lab 17: Endocrine Glands and Hormones

◆ Practice

1. The endocrine and nervous systems work together toward what purpose?

   Coordinating cellular function

2. Compare the endocrine and nervous systems by filling in the following table:

	Endocrine System	Nervous System
Chemical Messenger	blood	synapse
Site of Action (local/widespread)	Widespread	local
Speed of Response	slow	fast
Duration of Response	prolonged	short

3. Select one or more of the key choices below for each of the endocrine glands listed:

   a. Growth Hormone Releasing Hormone
   b. Oxytocin
   c. Prolactin
   d. Melatonin
   e. Thyroid stimulating hormone
   f. Glucagon
   g. Antidiuretic hormone
   h. Epinephrine
   i. Insulin
   j. ACTH
   k. Estrogen
   l. Testosterone
   m. Aldosterone
   n. Progesterone
   o. Thymosin

   Which hormone(s) does each endocrine gland **synthesize?**

   _____ Ovaries

   _____ Testes

   _H,_ _____ Adrenal medulla

   _____ Pineal gland

   _____ Adrenal cortex

   _I, F_ Pancreas

   _____ Adenohypophysis

   _____ Hypothalamus

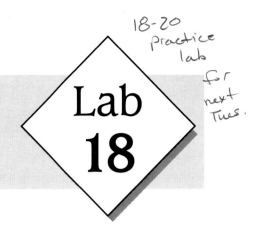

COMPONENTS OF WHOLE BLOOD

Lab 18

Objectives

1. Identify the different kinds of formed elements in the blood
2. Explain the diagnostic significance of the hematocrit
3. Explain the significance and conduct a differential WBC count
4. Describe the effects on formed elements of infectious mononucleosis, eosinophilia and lymphocytic leukemia

Materials

- Sheep blood
- Centrifuge
- Pipets
- Centrifuge tubes
- Microscopes
- Microscope slides
  - Whole blood smear
  - Infectious mononucleosis
  - Eosinophilia
  - Lymphocytic leukemia

INTRODUCTION

Blood appears to be a homogenous red fluid. Closer analysis, however, reveals that the blood has two major components, formed elements and plasma. **Formed elements** make up about 45% of the blood and can further be subdivided into **erythrocytes** (red blood cells or RBCs), **leukocytes** (white blood cells or WBCs), and **platelets** (thrombocytes). Together leukocytes and platelets make up less than one percent of whole blood. **Plasma** constitutes approximately 55% of the blood volume and contains water, lipids, dissolved substances, colloidal proteins and clotting factors. The study of blood, termed hematology, is important not only as part of the cardiovascular system but because it has significant clinical implications. Changes in the numbers and types of blood cells may be used as indicators of diseases.

CONCEPTS

I. The Composition of Whole Blood

A centrifuge is used to separate the components of blood for visualization and analysis. Blood is placed in tubes which are positioned opposite one another and angled outward on a wheel. The centrifuge rapidly spins the wheel and tubes. Heavier components within the fluid are pulled to the bottom of the tube by gravity, leaving the lighter components at the top.

◆ Exercise 18.1   *Examine the Components of Centrifuged Sheep Blood*

1. Obtain a small amount of sheep blood and use a transfer pipet to divide it equally between two centrifuge tubes.
2. Place the centrifuge tubes directly across from one another in the centrifuge. This keeps the centrifuge balanced while it spins at high speed.
3. Your instructor will set the centrifuge to a preprogrammed speed and time. Start the centrifuge and allow it to complete its program.
4. Remove the centrifuge tubes and observe the different layers within whole blood. The heaviest elements, the erythrocytes, spin to the bottom of the tube and the lighter plasma remains at the top. A very thin white line between the RBCs and plasma is the buffy coat, containing all of the WBCs and platelets.

     The percentage of RBCs in whole blood is known as the **hematocrit** (Hct), and is a diagnostically important value. The Hct is determined by the height of the RBCs in the centrifuge tube compared with the height of the total column of blood. The average Hct values for male and female respectively are 46% and 41%. The Hct may fall as low as 15% in severe anemia or rise to as high as 70% in polycythemia.

◆◆◆

II. Formed Elements of Blood

A. Function and Identification of Formed Elements

Erythrocytes, leukocytes and platelets all have unique properties and functions. Erythrocytes function to transport respiratory gases within the body. Leukocytes are important agents of the body's defense system against invasion and infection by foreign antigens or objects. WBCs are also scavengers, cleaning up debris and dead tissue. The two classes of WBCs can be distinguished in the microscope: **granulocytes,** which contain visible cytoplasmic granules, and **agranulocytes,** which do not contain cytoplasmic granules. Platelets are instrumental in stopping the flow of blood through tears in blood vessels.

◆ Exercise 18.2   *Identify Formed Elements on a Microscope Slide*

In this experiment, you will distinguish between RBCs and WBCs while examining professionally prepared slides of human blood.

1. Using Photos 377a–381c and lab charts as a reference, examine microscope slides of whole blood smears. Start at low power then view the slides at 40X and 100X.
2. Identify red blood cells, eosinophils, neutrophils, basophils, monocytes, and lymphocytes.

◆◆◆

Table 18.1
Formed Elements

◆◆◆

Formed Element	Structure and Appearance	Cells/mm³ of Blood	Function
Erythrocytes	Biconcave, anucleate, stains pink	4–6 million	Contains millions of hemoglobin molecules to transport oxygen and carbon dioxide
Leukocytes		5,000–11,000	
Granulocytes			
Neutrophils	Multilobed nucleus, faint cytoplasmic granules, stains purple	3000–7000	First WBC type to arrive at site of infection, phagocytizes bacteria
Eosinophils	Bilobed nucleus, orange stained cytoplasmic granules	100–400	Ejects digestive enzyme onto parasitic worms
Basophils	Purple stained multilobed nucleus, large purple stained cytoplasmic granules	20–50	Releases chemical mediators of inflammation (including histamine)
Agranulocytes			
Lymphocytes	Large, round purple nucleus	1500–3000	T and B lymphocytes recognize and inactivate antigens or infected body cells
Monocytes	Purple U-shaped nucleus, larger than RBCs	100–700	Becomes macrophage in tissue, phagocytizes antigens, dead or infected body cells, foreign matter
Platelets	Small purple dots are cytoplasmic fragments, anucleate	150,000–400,000	Releases chemicals that increase vascular spasm and attracts and agglutinates platelets to seal tears in blood vessels

B. Differential White Blood Cell Counts

White blood cells are the agents of the body's defense system. Each white blood cell type has specific protective functions. Elevated percentages of specific WBC types may be indicative of a pathological condition, therefore differential WBC counts are useful in diagnosing illness. In the following exercise you will examine blood smears from healthy individuals and from patients with infectious mononucleosis, eosinophilia and leukemia.

Infectious Mononucleosis is a disease caused by the Epstein-Barr virus that causes symptoms similar to the common cold. Significantly elevated percentages of atypical agranulocytes are a sign of this highly contagious condition. Though there is no cure for "Mono," the patient will usually recover within a few weeks.

Eosinophilia is a condition of elevated percentages of eosinophils, often in response to a parasitic infection of helminths, a microscopic worm.

Lymphocytic Leukemia is a cancerous condition of lymphocytes. Undifferentiated stem cells continue to divide without producing functional lymphocytes. The bone marrow and systemic circulation show increased numbers of dysfunctional lymphocytic cells, and the immune system becomes severely immunosuppressed.

◆ **Exercise 18.3** *Perform Differential White Blood Cell Counts*

Working in groups of four, perform a differential WBC count by counting cells at 40X objective magnification. Count 10 different microscopic fields of view by picking a landmark in each field of view and moving the microscope slide to a new field of view 10 times. For counts above 100, write "too numerous to count." Each student should count WBCs in one of the following microscope slides: normal blood, infectious mononucleosis, eosinophilia and lymphocytic leukemia.

Record class data here. Write and circle the average count for each cell type.

Cell Count	Normal Blood	Infectious Mononucleosis	Eosinophilia	Leukemia
Eosinophils				
Neutrophils				
Basophils				
Monocytes				
Lymphocytes				

Do the results agree with the above descriptions of each illness? _____

Name _____

Lab 18: Components of Whole Blood

◆ Practice

1. In the following table, briefly describe the structural characteristics and functions of each component of blood.

Components	Structural Characteristics	Functions
RBCs		
Basophil		
Eosinophil		
Neutrophil		
Lymphocyte		
Monocyte		
Platelets		

2. What is the clinical significance of the RBC count?

3. What factors besides gender might affect the RBC count? *(Hint: Think about RBC function.)*

4. If you had a high Hct, would you expect your blood hemoglobin concentration to be high? Why?

5. Carbon monoxide, present in automobile exhaust fumes, binds more tightly to hemoglobin than oxygen. Explain why inhalation of CO is life threatening.

BLOOD GROUPS

Lab
19

Objectives

1. Define the terms **agglutinin** and **agglutinogen**
2. Describe what agglutinins and agglutinogens are present in each blood type
3. Explain what blood types are compatible for purposes of transfusion
4. Describe the role of the Rh factor in erythroblastosis fetalis

Materials

- Ward's blood type kit (Ward's # 36 W 0019; refill kit 36 W 0034)
- 3 well agglutination slides
- toothpicks

INTRODUCTION

Blood groups are determined by the presence of glycoproteins on the surface of red blood cells. These have significance when a blood transfusion is necessary, since some blood groups are incompatible with others. A blood transfusion often involves removing the plasma and buffy coat from whole blood to leave the red blood cells (packed cells). If incompatible packed cells are donated to a patient, blood clots (thromboses) can form within the patient's blood vessels and block blood flow to vital organs. In this lab you will investigate the physiological basis of blood groups and practice typing synthetic blood using slide agglutination tests.

CONCEPTS

I. ABO Blood Groups

A person's blood type (blood group) depends on the presence of specific **glycoproteins** on the surface of red blood cell plasma membranes. These glycoproteins have the potential to react with antibodies present in the blood plasma. Such a reaction **agglutinates,** or clumps, the proteins and attached RBCs together, causing the formation of a blood clot **(thrombosis).** We call the glycoproteins **agglutinogens** because of their ability to act like an antigen in binding to antibodies.

There are two significant classes of agglutinogens that contribute to the ABO blood types: **A** and **B.** A person's red blood cells may have either, both or neither class. Individuals with only A agglutinogens in their RBCs have blood type A. Those with only B agglutinogens have blood type B. A person with type AB blood has RBCs containing both A and B agglutinogens. Some individuals have neither A nor B agglutinogens and are classified as type O. Under normal physiological conditions a person's blood plasma does not contain antibodies to the agglutinogens present in her RBCs, and blood agglutination does not occur.

Caution must be observed when transfusing whole blood or packed RBCs because of a unique property of blood plasma antibodies. Although plasma does not normally contain antibodies to agglutinogens present on RBCs, antibodies to *foreign* agglutinogens are present. This is true even though the blood has never been directly exposed to foreign agglutinogens. (In most other situations, blood must be exposed to an antigen prior to making antibodies specific to that antigen.) Antibodies to A and B agglutinogens are termed **agglutinins.** Thus, an individual with type A blood has no agglutinins which bind type A agglutinogens, but does possess type B agglutinins. Should this person receive a transfusion of RBCs with B agglutinogens, her B agglutinins will bind to the foreign antigen and will cause agglutination.

◆ **Exercise 19.1** *List Agglutinogens and Agglutinins Present for Each Blood Group*

Test your understanding of the presence of agglutinogens and agglutinins in ABO blood groups by filling in Table 19.1.

Table 19.1
Agglutinogens and Agglutinins in Blood Groups

ABO Blood Group	A	B	AB	O
Agglutinogens present				
Agglutinins present				

II. Rh Factor

Another agglutinogen present in some people is the **Rh factor,** named for its discovery in the Rhesus monkey. Individuals with the Rh factor are said to be **Rh positive** (Rh⁺), and those without are **Rh negative** (Rh⁻). Any ABO blood group can be Rh⁺ or Rh⁻. Unlike the A and B agglutinogens, however, anti-Rh antibodies are not present in the blood plasma until after the plasma has been exposed to the Rh antigen.

Health professionals must match Rh⁻ packed cell transfusions with Rh⁻ patients. Because Rh⁺ individuals will never make antibodies against the Rh factor, they can receive Rh⁺ or Rh⁻ blood transfusions.

Erythroblastosis Fetalis

Pregnant women who are Rh⁻ also need to be aware of the role of the Rh factor. Because blood cells do not cross the placenta an Rh⁻ mother will not make antibodies against the Rh factor if she carries an Rh+ fetus. However, during birth the placenta pulls away from the mother's uterus causing some bleeding. The fetal blood is mixed with the mother's blood, and she will begin to manufacture anti-Rh antibodies in her plasma. This becomes significant if she becomes pregnant with another Rh⁺ fetus. The anti-Rh antibodies are able to cross the placenta, and will bind to the Rh glycoproteins on the fetal red blood cells and cause agglutination. The resulting condition in the fetus is called **erythroblastosis fetalis,** and creates anemia and hypoxia within the baby. If untreated, the baby may experience brain damage or death. Newborns, and even fetuses in the uterus, can receive a blood transfusion to provide additional red blood cells for oxygen transport.

To prevent the occurrence of erythroblastosis fetalis, pregnant Rh⁻ women are given an injection of **RhoGAM,** an agglutinin that binds anti-Rh antibodies. RhoGAM agglutinates the anti-Rh antibodies in the mother and prevents their transmission to the fetus. The woman will receive RhoGAM each time she becomes pregnant.

◆ Exercise 19.2 *Practice Using Slide Agglutination Tests to Type Blood*

In this lab exercise you will use artificial blood to perform a slide agglutination test in which you will determine the blood types of four fictional individuals. Because the antibodies in this test are designed to work with specially prepared artificial blood, they will not work on real blood to determine your own blood type.

1. Obtain the following supplies:

   toothpicks
   four blood typing well slides
   artificial blood from four different fictional individuals
   vials of anti-A, anti-B, and anti-Rh antibodies

2. Use a separate slide for each individual's "blood." In the well labeled "A" place one or two drops of the anti-A antibodies. Put one or two drops of anti-B antibodies in the "B" well and anti-Rh antibodies in the Rh well.

3. Place a drop of "blood" from one individual in each of the labeled wells and mix using toothpicks. Do not use the same end of the toothpick in more than one well or you will contaminate the solutions and invalidate your results.

4. Observe the wells for agglutination. If agglutination occurs, this indicates that the antibodies are binding with a specific antigen.

   If agglutination occurs in the "A" well, are "A" agglutinogens present or not? _____.

   Record your results in Table 19.2.

5. Repeat the test for the remaining three "blood" samples.

Table 19.2
Blood Type Results

◆◆◆

Patient Name	Anti-A Result	Anti-B Result	Anti-Rh Result	Blood Type

◆◆◆

Name _____

Lab 19: Blood Groups

 Practice

1. Define *agglutinogen:*

2. Define *agglutinin:*

3. From memory, list the agglutinogens and agglutinins present in each blood group.

4. Why is it dangerous to transfuse incompatible packed cells into a patient?

5. Does it matter whether an Rh⁺ patient receives Rh⁺ or Rh⁻ blood? Why?

6. Does it matter whether an Rh⁻ patient receives Rh⁺ or Rh⁻ blood? Why?

7. What are the implications of an Rh⁺ mother pregnant with an Rh⁻ fetus?

8. What is the significance of an Rh⁻ mother pregnant for the second time with an Rh⁺ baby?

9. How is erythroblastosis fetalis prevented? What is the mechanism of treatment?

ANATOMY OF THE HEART

Lab 20

Objectives

1. Identify specific structures in the heart
2. Identify the great vessels entering and exiting the heart
3. Trace the pathway of blood through the heart and great vessels
4. Identify blood vessels that supply the heart

Materials

- Model of human heart
- Sheep heart
- Disposable gloves
- Dissecting instruments

INTRODUCTION

The heart has fascinated people for as long as its existence has been known. While we know this not to be true, the heart is often credited for a person's *feelings,* or *gut reaction.* Matters of central significance are said to be "at the heart of the matter."

Anatomically, the human heart is merely a pump that moves blood through two main circulatory systems. It rests in the mediastinum intermediate to the right and left lungs. The **diaphragmatic surface** rests on the diaphragm muscle. Pointing to the left and somewhat inferiorly is the **apex.** The superoposterior surface, where the great vessels join with the heart, is called the **base.** Internally the heart contains four chambers, the superior **atria** which receive blood, and the inferior **ventricles** which pump blood out. In this lab we will examine the great vessels that deliver blood directly to or from the heart, and the external structures and internal structures of the heart.

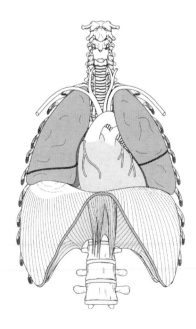

FIGURE 20.1 ◆ Position of the Heart

CONCEPTS

I. Great Vessels

The great vessels are large diameter, elastic and thick muscular vessels that are attached to the heart. **Arteries** are vessels that carry blood away from the heart, and **veins** carry blood to the heart. All great vessels connect to the heart on the superior or superoposterior surface.

The **superior vena cava** delivers blood to the heart from the head and upper extremities. The **inferior vena cava** delivers blood to the heart from the thorax, abdominopelvic region and lower extremities. Both empty into the right atrium.

The **ascending aorta** exits superiorly and curves posteriorly as the **aortic arch** to descend posterior to the heart and lungs in the thoracic cavity. Blood exits the left ventricle of the heart and enters the aorta to be delivered to all systemic regions. Branching off of the aortic arch are three large vessels that supply the head and upper extremities. The first branch as blood is leaving the heart is the unpaired **brachiocephalic artery or trunk,** which delivers blood to the right upper extremity and right head regions. The left **common carotid** artery is the middle branch, which carries blood to the left head region. The leftmost branch is the left **subclavian** artery, which sends blood to the left upper extremity.

Immediately to the left of the ascending aorta exits the large **pulmonary trunk** from the right ventricle. This artery bifurcates into the **left pulmonary artery,** which serves the left lung, and the **right pulmonary artery,** which travels inferior to the aortic arch to serve the right lung. Posteriorly, two pairs of **pulmonary veins,** one pair from each lung, return blood to the left atrium.

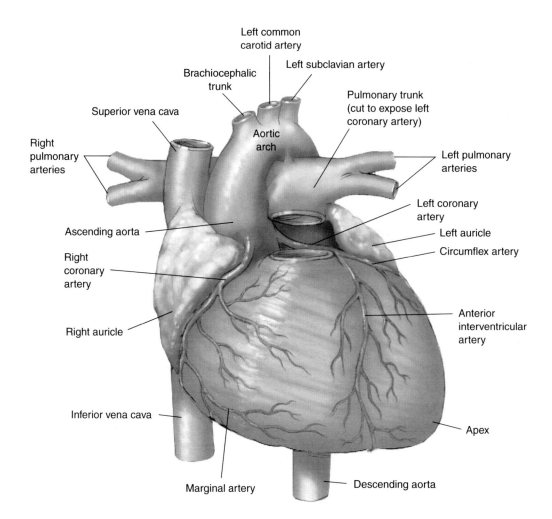

FIGURE 20.2 ◆ Anterior View of the Heart

II. Surface Structures of the Heart

Externally the heart is divided into a superior and inferior portion by the **atrioventricular groove** (or sulcus). This groove is sometimes called the coronary sulcus. Superior to this groove are the **auricles,** within which are the atria. Inferior to the atrioventricular groove, as the name suggests, are the ventricles. Adhered to the external surface of the heart is the visceral pericardium, also known as the **epicardium.** The epicardium is considered the outer layer of the heart wall. On the anterior surface a groove extends roughly vertically from the atrioventricular groove to the diaphragmatic surface. This vertical groove is located superficial to the septum between the two ventricles, and so is called the **anterior interventricular** groove. Posteriorly the **posterior interventricular groove** mirrors the anterior surface.

Connecting the pulmonary trunk to the aortic arch is a thin, tendinous **ligamentum arteriosum.** This structure is a remnant of a fetal duct, the **ductus arteriosus,** that allowed blood to pass from the pulmonary trunk into the aorta, thus bypassing the lungs.

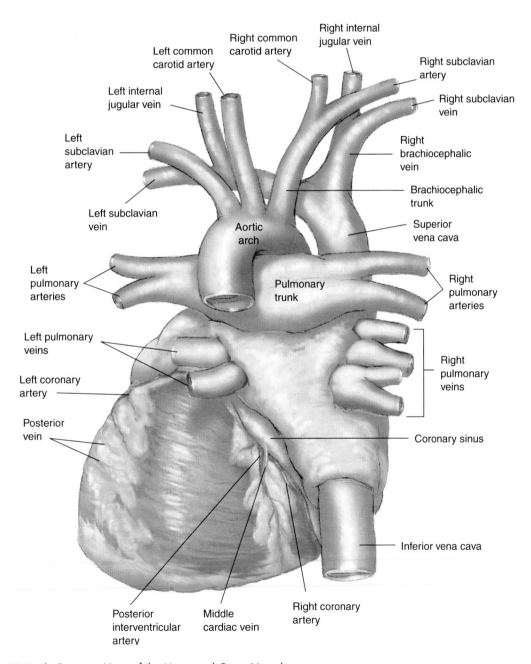

FIGURE 20.3 ◆ Posterior View of the Heart and Great Vessels

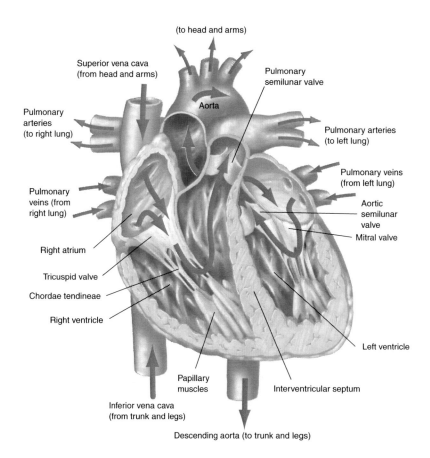

(to head and arms)

Superior vena cava
(from head and arms)

Pulmonary
semilunar valve

Aorta

Pulmonary
arteries
(to right lung)

Pulmonary arteries
(to left lung)

Pulmonary veins
(from left lung)

Pulmonary
veins (from
right lung)

Aortic
semilunar
valve

Mitral valve

Right atrium

Tricuspid valve

Chordae tendineae

Right ventricle

Left ventricle

Papillary
muscles

Interventricular septum

Inferior vena cava
(from trunk and legs)

Descending aorta (to trunk and legs)

FIGURE 20.4 ◆ Internal Structures and Blood Flow through Heart

III. Blood Supply to the Heart

The heart pumps blood to all of the body's tissues, including heart tissue. The coronary arteries and cardiac veins are usually encased in fat. The right and left **coronary arteries** arise from the ascending aorta at the atrioventricular groove. The coronary arteries supply blood to the myocardium, or muscular layer of the heart wall. The left coronary artery divides into two branches posterior to the pulmonary trunk: the **anterior interventricular artery** travels in the groove of the same name, and the **circumflex artery** rests in the left atrioventricular groove. The right coronary artery is positioned in the right atrioventricular groove. Its branches are the **posterior interventricular artery** in the groove of the same name, and the **marginal artery** on the right side of the heart.

Three cardiac veins drain blood from the heart into the **coronary sinus,** located in the posterior atrioventricular groove. The **great cardiac vein** rests in the anterior interventricular groove, the **middle cardiac vein** is located in the posterior interventricular groove, and the **small cardiac vein** is beside the marginal artery. The coronary sinus empties blood into the right atrium.

IV. Internal Heart Structures

The atrial walls are lined with ridged **pectinate muscles.** Separating the two atria is the **interatrial septum.** Within the interatrial septum, visible on the right side, is a shallow depression called the **fossa ovalis.** (This structure is a remnant of a fetal passageway, the **foramen ovale,** from the right atrium to the left, allowing blood to bypass pulmonary circulation.) **Atrioventricular valves** allow blood to passively drain into the ventricles below. The right atrioventricular (AV) valve is also called the **tricuspid** because of the three flaps that comprise the valve. The left AV valve is also known as the **bicuspid,** for its two flaps, and the **mitral** valve. The AV valves are secured to finger-like muscle projections within the ventricular walls, called **papillary muscles,** by fibrous **chordae tendinae.** Other muscular ridges of the ventricular walls are the **trabeculae carnae.** Dividing the two ventricles is the **interventricular septum.** Blood exits the right ventricle superiorly through the pulmonary trunk, and the left ventricle through the ascending aorta. Both the **pulmonary** and **aortic semilunar valves** prevent the backflow of blood into the ventricles. Note that the thickness of the left ventricular wall is much greater than that of the right side. Why do you think this is so?

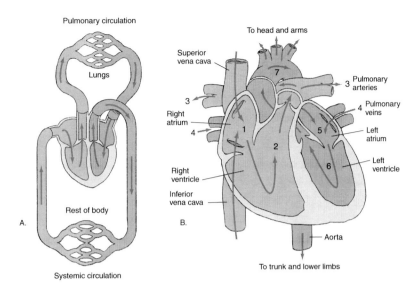

FIGURE 20.5 ◆ Pulmonary and Systemic Circulation

V. Circulation

The heart consists of two pumps, each of which serves a separate system of circulation. **Systemic circulation** involves the movement of oxygen rich, carbon dioxide poor blood from the heart to all of the body tissues, and the return of oxygen poor, carbon dioxide rich blood back to the heart. **Pulmonary circulation** involves the movement of oxygen poor, carbon dioxide rich blood from the heart to the lungs, and oxygen rich, carbon dioxide poor blood from the lungs back to the heart. Each side of the heart serves one circulation system.

The right side of the heart pumps blood from the right ventricle into **pulmonary circulation.** Blood moves from the right ventricle, through the pulmonary semilunar valve, into the pulmonary trunk. Blood continues into the right and left pulmonary arteries and ultimately into the pulmonary capillaries surrounding the lungs. Here gas exchange occurs, loading oxygen into the capillary blood and unloading carbon dioxide into the lungs. The oxygen rich blood travels from the pulmonary capillaries into the pulmonary veins and back to the left atrium.

From the left atrium, blood drains through the left AV valve into the left ventricle. The left side of the heart serves the **systemic circulation.** The left ventricle pumps blood through the aortic semilunar valve into the ascending aorta. Blood exits the aorta through many muscular arteries to be delivered to capillary beds throughout the body. At systemic capillaries, oxygen is off loaded at body tissues and carbon dioxide waste gas is picked up. The oxygen poor blood is returned via veins to the superior and inferior vena cavae, and ultimately to the right atrium. Blood then drains from the right atrium through the right AV valve into the right ventricle.

The left ventricular wall is much thicker than the right side because much greater pressure is required to pump blood from the heart to the farthest body tissues. The right pulmonary pump needs only pump blood as far as the adjacent lungs.

◆ Exercise 20.1   *Identify Structures in a Human Heart Model*

Trace the pathway of blood through a model of the human heart. Identify the following structures (Photos 382 and 383):
- Great vessels
  - pulmonary trunk, pulmonary arteries, ascending aorta, aortic arch, brachiocephalic trunk, left common carotid artery, left subclavian artery, superior vena cava, inferior vena cava, pulmonary veins
- External structures
  - auricles, atrioventricular groove (coronary sulcus), anterior and posterior interventricular grooves, coronary arteries, anterior and posterior interventricular arteries, marginal artery, great cardiac vein, middle cardiac vein, small cardiac vein, coronary sinus, ligamentum arteriosum
- Internal structures
  - Right and left atria, fossa ovalis, right and left AV valves, pectinate muscles, right and left ventricles, chordae tendinae, papillary muscles, trabeculae carnae, pulmonary and aortic semilunar valves

◆ Exercise 20.2 *Dissect and Identify Structures in a Sheep Heart*

We will use a sheep heart, which is similar in structure and function to a human heart, to identify external and internal structures. Use Photos 384a and 384b as a reference.

1. Obtain a sheep heart and observe the external surface. If the heart has an intact pericardium, cut open the parietal layer and observe the visceral pericardium (epicardium) on the surface of the heart.
2. Remove the pericardium. You may also need to remove fatty tissue that is attached to the great vessels and surface of the heart.
3. Identify the anterior and posterior surfaces of the heart. Great vessels exit out of the posterior side. Try to identify the great vessels. This will be more obvious once you have cut the heart open to expose the internal chambers.
4. Identify the **auricles,** the **atrioventricular groove,** and the **anterior** and **posterior interventricular grooves.**
   ▶ What blood vessels are present in each groove?

5. Cut the heart into anterior and posterior halves with a scalpel. Make your incision perpendicular to the anterior interventricular groove.
6. Identify the four chambers.
   ▶ How can you distinguish between the right and left sides of the heart?

7. Identify the following structures: **right** and **left AV valves, pulmonary** and **aortic semilunar valves, pectinate muscles, trabeculae carnae, chordae tendinae, papillary muscles**
8. Now that the chambers are exposed, confirm the identification of the great vessels: **pulmonary trunk, ascending aorta, superior** and **inferior vena cavae** (if present), opening for pulmonary veins

◆◆◆

Name _____

Lab 20: Anatomy of the Heart

◆ **Practice**

1. Starting with the superior and inferior vena cavae, list the structures through which blood flows through the heart. Include all valves. End with the aortic arch.

2. What is the function of the AV valves? What is the function of the semilunar valves? What would be the result of leaky valves?

3. What is the role of the chordae tendinae?

4. Describe the roles of pulmonary and systemic circulations.

5. What differences exist in the walls of the right and left ventricles? What is the reason for this difference?

6. Describe two structures that are remnants of fetal circulation. Why is fetal circulation different than that of a newborn baby?

7. Through which main arteries does the heart tissue receive blood? What is the result if one of these arteries is blocked?

8. Explain how the heart functions as a double pump.

ANATOMY OF BLOOD VESSELS

Lab 21

Objectives

1. Describe the structural and functional differences between arteries, veins and capillaries
2. Identify the major arteries and veins in human models and diagrams
3. Dissect and identify the major arteries and veins in a cat specimen
4. Name the blood vessels in and describe the functions of special circulations: pulmonary, hepatic portal, fetal

Materials

- Microscope slides of cross sections through arteries and veins
- Microscopes
- Models
  - Human circulatory system
  - Human torso, arm and leg
- Cat specimens
- Dissecting trays
- Dissecting instruments
- Disposable gloves

INTRODUCTION

Blood vessels are the body's transport mechanism for blood and its contents. As we learned in Lab 19, blood exits the heart through arteries and returns to the heart within veins. Connecting arteries and veins are thin walled capillaries. Gases, fluid and solutes enter and exit circulation from the capillaries. White blood cells also exit the blood through capillaries to protect tissues from infection.

Most blood vessels are named for the region of the body in which they are located, or the region that they supply. Just as a road may continue through several cities and have a different name in each city, a blood vessel may change names as branches arise or as it enters different locations. In this lab you will begin at the heart and identify major vessels in the body.

CONCEPTS

I. Structure and Function of Blood Vessels

A. Blood Vessel Walls

Both arteries and veins have three layers in the vessel wall. The innermost layer, in contact with blood within the lumen, is called the **tunica interna** (or intima). This layer is one cell thick and is composed of endothelium. The next deepest layer is the **tunica media,** which is composed of perpendicular sheets of smooth muscle. The longitudinal layer runs lengthwise along the vessel, while the circular layer encircles the vessel. When the circular layer contracts, the lumen of the vessel becomes smaller, causing **vasoconstriction.** Muscle relaxation allows **vasodilation** of the vessel. Vasomotor fibers, part of the sympathetic nervous system, innervate the smooth muscle and provide a constant low level of excitation to create vasomotor tone. The outermost layer is the **tunica externa** (or adventitia). Collagen fibers within the tunica externa provide strength and also serve to anchor the vessel to surrounding tissue.

Arteries and veins have all three layers in their walls, however capillaries are composed only of the tunica interna, or just one layer of endothelial tissue.

239

B. Arteries

There are actually three different types of arteries: elastic, muscular and arteriole. **Elastic arteries** contain a very thick tunica media which is partly composed of elastic fibers. These arteries, which are found closest to the heart, are able to withstand high blood pressures. They stretch to receive large volumes of blood, then recoil quickly to further propel blood distally along the vessel. **Muscular arteries** are similar to highways that provide routes from state to state; they deliver blood to specific organs in the body. Muscular arteries are also called distributing arteries. Their tunica media contains proportionately more smooth muscle and less elastic fibers than elastic arteries. Most of the arteries that you will identify in this lab are muscular arteries. **Arterioles** are the smallest arteries, and feed into capillaries. The smallest arterioles may consist only of a single layer of endothelium surrounded by a single layer of smooth muscle. Vasoconstriction and vasodilation of the arterioles determines blood flow into the capillaries, and thus how much oxygen, fluid, nutrients and other substances are available to tissues.

C. Capillaries

Capillaries are arranged in branching networks called capillary beds. Arterioles deliver blood into a straight **vascular shunt** that is continuous with a venule, or small vein. Branches exit from the first part of the vascular shunt, and are controlled with circular bands of smooth muscle called **precapillary sphincters.** The region of the vascular shunt that has precapillary sphincters is called the metarteriole. The tiny branches are **true capillaries,** and after branching and networking, they return blood to the venous end of the vascular shunt, or thoroughfare channel. Tissue needs determine the activity of the precapillary sphincters, and the **perfusion** of blood to tissues within the true capillaries. Greater perfusion allows greater gas exchange, fluid leakage and solute delivery.

D. Veins

Venules are the smallest veins, which receive blood from the vascular shunt of the capillary bed. The smallest venules consist of only endothelium and a layer or two of smooth muscle, while larger venules have all three wall layers. Venules feed into larger **veins.** The lumen within veins is much larger than in corresponding arteries. Although veins have all three tunics in their walls, the characteristics are different than in arteries. The tunica interna invaginates at regular intervals within vessels of the limbs to form **venous valves.** The valves allow blood to flow distally within the vein, but prevent backflow which would otherwise occur due to gravity and low blood pressure. Thinner layers of smooth muscle within the tunica media result in "floppier" walls, and give the veins a collapsed appearance. The tunica externa is thicker in veins than in arteries, and consists of bundles of longitudinally oriented collagen fibers.

A larger total volume is present in the venous system than in the arterial system. 65% of the body's total blood supply is present within the veins at all times, giving the venous system the name **venous reservoir,** or **blood reservoir.**

◆ Exercise 21.1   *Examine Arteries and Veins under a Microscope*

Examine a microscope slide with arteries and veins at low power under a microscope. Use Photo 385b as a reference. Draw the three layers of the vessel walls and the shape of the vessel lumens in the space below.

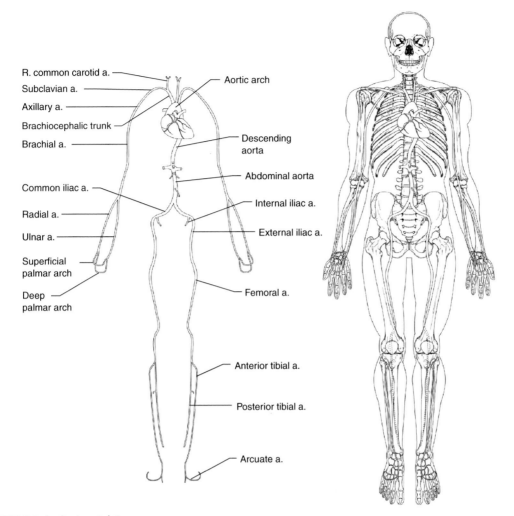

R. common carotid a.

Subclavian a.

Axillary a.

Brachiocephalic trunk

Brachial a.

Aortic arch

Descending aorta

Abdominal aorta

Common iliac a.

Radial a.

Ulnar a.

Superficial palmar arch

Deep palmar arch

Internal iliac a.

External iliac a.

Femoral a.

Anterior tibial a.

Posterior tibial a.

Arcuate a.

FIGURE 21.1 ◆ Arterial System

II. Identification of Major Blood Vessels

 Exercise 21.2 *Identify Arteries in Diagrams and Models*

Arteries tend to be located in deep, well protected areas in the body. Use the figures in this lab and Photo 385a to help you identify the following arteries and their branches on pictures and models. Begin at the heart and work distally. Take note of which arteries are paired (are present on the right and left) and unpaired.

◆◆◆

A. **Arteries of the head, neck, and thorax (Figures 21.1, 21.2, and 21.5)**

   The aortic arch gives off three main branches, each of which subsequently branches several times.

   1. **Ascending Aorta** (unpaired)
   2. **Arch of the Aorta** (unpaired) branches
      a) **Brachiocephalic trunk** (unpaired)
         1) **Right common carotid** a.
            • Right **internal carotid** a. (supplies orbits and brain)
            • Right **external carotid** a. (supplies head)

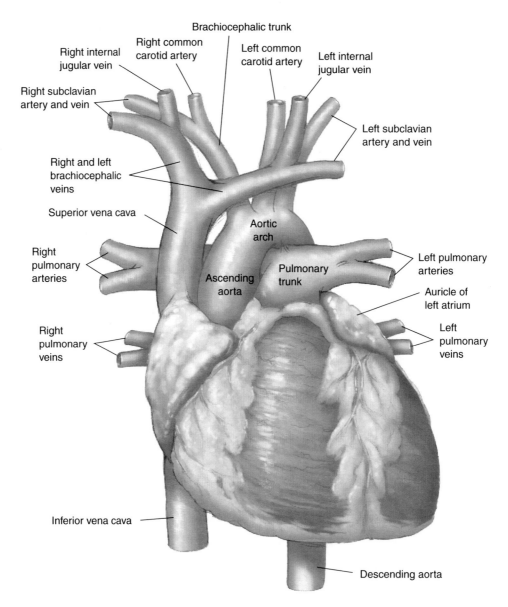

FIGURE 21.2 ◆ Great Vessels and Their Branches

    2) Right subclavian a.
-     Right **vertebral** a. (supplies vertebra, spinal cord, brain)
-     Right **internal thoracic** a. (supplies anterior thoracic wall, skin, mammary glands)

  b) Left common carotid a.
    1) Left **internal carotid** a. (supplies orbits and brain)
    2) Left **external carotid** a. (supplies head)

  c) Left subclavian a.
    1) Left **vertebral** a. (supplies vertebra, spinal cord, brain)
    2) Left **internal thoracic** a. (supplies anterior thoracic wall, skin, mammary glands)

3. Descending aorta (unpaired) descends in the thorax posterior to the heart and lungs
  a) Posterior intercostal arteries (supply intercostals muscles, vertebrae, spinal cord) exit at intervals down the length of the descending aorta

B. Arteries of the upper limbs (Figures 21.3 and 21.4)
These arteries are all paired, one on each side.
1. Subclavian a. becomes axillary a.
2. Axillary a. becomes brachial a. (after deep brachial a. branches off)

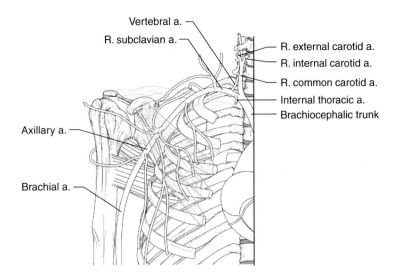

FIGURE 21.3 ◆ Arteries of the Shoulder

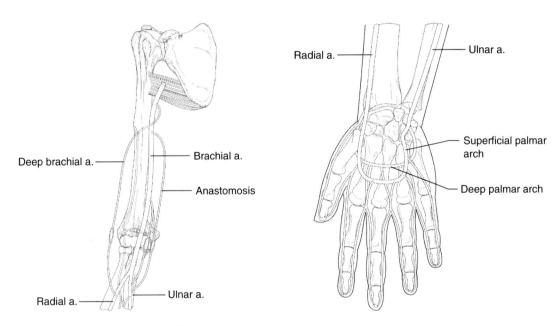

FIGURE 21.4 ◆ Arteries of the Arm and Hand

   3. **Brachial** a. (supplies arm flexors) branches
      a) **Radial** a. (supplies lateral forearm, wrist, thumb)
      b) **Ulnar** a. (supplies medial forearm)
         1) **Superficial and deep palmar arches** are continuous with the radial and ulnar arteries (supply fingers)
C. **Arteries of the abdomen and pelvis (Figure 21.5)**
   The descending aorta passes through a foramen in the diaphragm muscle, the aortic hiatus, to enter the abdomen as the abdominal aorta. Branches off the abdominal aorta serve abdominal organs. Branches are listed in the order that they exit the abdominal aorta. At the level of L_4 the abdominal aorta divides into the right and left common iliac arteries.
   1. **Abdominal Aorta** (unpaired)
      a) **Celiac trunk** (unpaired)
         1) **Gastric** a. (supplies the stomach and inferior esophagus)
         2) **Common hepatic** a. (supplies the stomach, duodenum, pancreas and liver)
         3) **Splenic** a. (supplies the stomach, pancreas and spleen)

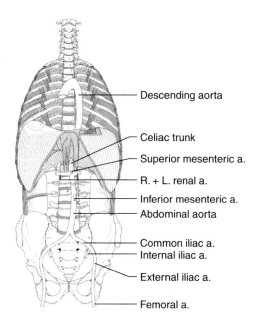

Descending aorta

Celiac trunk

Superior mesenteric a.

R. + L. renal a.

Inferior mesenteric a.

Abdominal aorta

Common iliac a.
Internal iliac a.

External iliac a.

Femoral a.

FIGURE 21.5 ◆ Arteries of the Abdomen

    b) Suprarenal a. (paired; supply the adrenal glands)
    c) Superior mesenteric a. (unpaired; supplies the small intestines via the intestinal arteries)
    d) Renal a. (paired; supply the kidneys)
    e) Gonadal a. (paired; supply the ovaries in females and testes in males)
    f) Inferior mesenteric a. (unpaired; supplies the distal large intestine)
  2. Common iliac a. (paired)
  The abdominal aorta splits into the right and left common iliac arteries
    a) Internal Iliac a. (paired; supply pelvic wall, bladder, rectum, gluteal muscles, thigh adductors, and female reproductive organs (uterus, vagina) or male prostate gland and ductus deferens)
  3. External iliac a. (paired continuation of common iliac a; supply anterior abdominal wall)
D. Arteries of the lower limb (Figures 21.6 and 21.7)
The external iliac a. passes through the inguinal canal in the pelvic floor to become the femoral a. All lower limb arteries are paired.
  1. Femoral a. branches
    a) Deep femoral a. (= profunda femoris a.; supplies the hamstrings, quadriceps and adductors)
    b) Popliteal a. is a continuation of the femoral a. as it passes through the adductor hiatus into the popliteal region (supplies knee structures) branches
      1) Anterior tibial a. (supplies extensor muscles of the leg)
       • **Dorsalis pedis** a. (supplies the foot)
      2) Posterior tibial a. (supplies flexor muscles of the leg)
       • **Fibular** a. (**peroneal** a.; supplies lateral leg muscles)

◆ **Exercise 21.3** *Identify Veins in Diagrams and Models*

In general, veins are located in association with arteries and take the same name as the artery with which they run. Some superficial veins do not have an associated artery. It is these superficial veins into which injections may be made and from which blood may be drawn. Blood enters veins at their capillary ends, and travels distally toward the heart. All venous blood eventually empties into either the superior or inferior vena cava to be delivered to the right atrium. Use the figures in this lab and Photo 385a to help you identify the following veins on pictures and models.

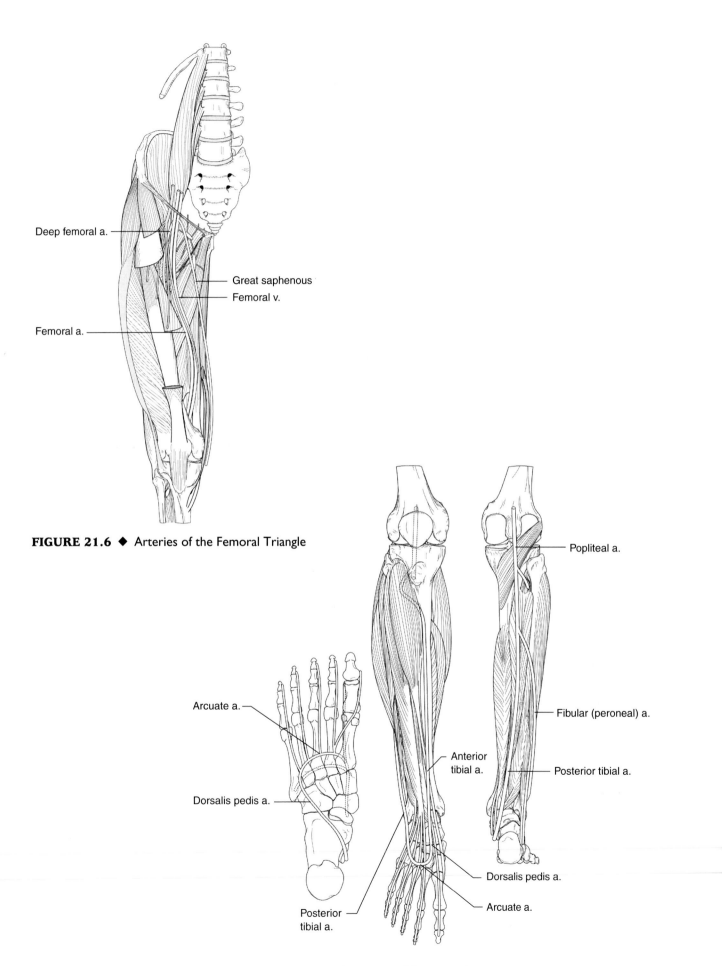

Deep femoral a.

Great saphenous

Femoral v.

Femoral a.

FIGURE 21.6 ◆ Arteries of the Femoral Triangle

Popliteal a.

Arcuate a.

Dorsalis pedis a.

Fibular (peroneal) a.

Anterior tibial a.

Posterior tibial a.

Posterior tibial a.

Dorsalis pedis a.

Arcuate a.

FIGURE 21.7 ◆ Arteries of the Leg and Foot

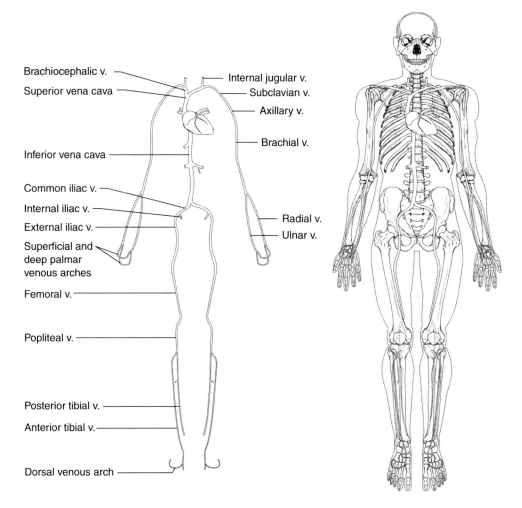

Brachiocephalic v.
Superior vena cava
Inferior vena cava
Common iliac v.
Internal iliac v.
External iliac v.
Superficial and
deep palmar
venous arches
Femoral v.
Popliteal v.
Posterior tibial v.
Anterior tibial v.
Dorsal venous arch

Internal jugular v.
Subclavian v.
Axillary v.
Brachial v.
Radial v.
Ulnar v.

FIGURE 21.8 ◆ Venous System (external jugular and vertebral veins not pictured)

E. **Veins draining the head, neck, and upper limb regions (Figures 21.2, 21.8, 21.9, and 21.10)**

The head, neck and upper limbs are drained by veins that empty into the **superior vena cava.** This list begins at the superior vena cava and follows the vessels against the flow of blood, away from the heart. All veins in this region are paired. Superior vena cava receives venous blood from:

1. **Brachiocephalic** v. (drains head and upper limb)
    a) **Subclavian** v. enters brachiocephalic
       Head region branches:
       1) **Internal jugular** v. (drains the brain)
       2) **Vertebral** v. (drain vertebrae, spinal cord, neck muscles)
       3) **External jugular** v. (drain head and face structures)
    Upper limb branches:
    b) **Axillary** v. enters subclavian v.
       1) **Cephalic** v. (drains lateral forearm and brachium)
       2) **Brachial** v. supplied by:
          • **Radial** v. (drains lateral forearm)
          • **Ulnar** v. (drains medial forearm)
          • **Superficial** and **deep palmer arches** (drain hand into radial and ulnar veins)
       3) **Basilic** v. (drains posteromedial forearm)
       4) **Median cubital** v. (connects cephalic v. to basilic v. in the cubital region, common site from which to draw blood for diagnostic purposes)

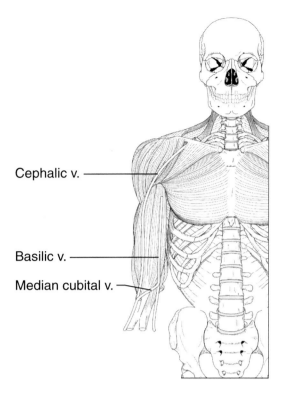

FIGURE 21.9 ◆ Superficial Veins of the Arm

Cephalic v.

Basilic v.

Median cubital v.

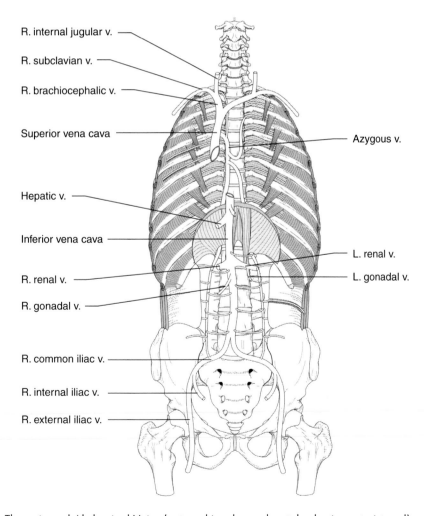

FIGURE 21.10 ◆ Thoracic and Abdominal Veins (external jugular and vertebral veins not pictured)

R. internal jugular v.

R. subclavian v.

R. brachiocephalic v.

Superior vena cava

Azygous v.

Hepatic v.

Inferior vena cava

R. renal v.

L. renal v.

R. gonadal v.

L. gonadal v.

R. common iliac v.

R. internal iliac v.

R. external iliac v.

F. **Veins draining the lower limbs and abdominopelvic regions (Figures 21.6 and 21.10)**
The **inferior vena cava** receives blood drained from the abdominal, pelvic, and lower limb regions.
  1. **Hepatic** v. (paired; drains liver)
     a) **Hepatic portal** v. receives blood from vessels which drain digestive viscera
        1) **Superior mesenteric** v.
        2) **Splenic** v.
        3) **Inferior mesenteric** v.
  2. **Right suprarenal** v. (drains right adrenal gland; left suprarenal v. drains into left renal v.)
  3. **Renal** v. (paired; drain kidneys; left renal v. has two branches) Left renal v. branches:
     a) **Left suprarenal** v. (drains left adrenal gland)
     b) **Left gonadal** v. (drains left ovary in females or left testis in males)
  4. **Right gonadal** v. (drains right ovary in females or right testis in males)
  5. **Common Iliac** v.
     a) **Internal iliac** v.
     b) **External iliac** v. (continuous with common iliac v.)
     Lower limb:
     c) **Femoral** v. (continuous with external iliac v., changes its name as it passes through the inguinal canal)
        1) **Great saphenous** v. (superficial vein courses medially down thigh and leg to join the **dorsalis pedis** v.)
     d) **Popliteal** v. (continuous with femoral v., changes its name as it passes through the adductor hiatus)
        1) **Anterior tibial** v.
        2) **Posterior tibial** v.
           • **Fibular** v. (**peroneal v.;** branches off of the posterior tibial v.)
           • **Dorsalis pedis** v. (drains the foot into the anterior and posterior tibial v.)

◆ Exercise 21.4 *Dissect and Identify Major Blood Vessels in a Cat Specimen*

In this exercise you will identify specific blood vessels in the cat. Recall that the vessels have been injected with blue (veins) and red (arteries) latex for ease of identification. This process artificially increases the size of the veins, but not the arteries. (Why?) Also, the thick walls of the elastic and some muscular arteries prevent the red latex from showing through, so these vessels may appear tan or gray.

Be careful not to cut vessels that you are required to know. Use Photos 411–413 and Figures 21.11 and 21.12 to help you identify the listed vessels. Do not remove or cut any organs at this time without direction from your instructor!

Arteries

1. Begin your dissection at the heart. Remove the pericardium from the heart and the great vessels. Identify the **ascending aorta** and the **aortic arch.** Follow the aortic arch as it curves to the left, posterior to the heart, to continue inferiorly as the **descending aorta.** Observe the tiny **posterior intercostal arteries** where they exit the descending aorta in the thoracic cavity.

2. Carefully use a blunt probe to identify the *red* pulmonary veins and *blue* pulmonary arteries where they enter the lungs.

   ▶ Why are pulmonary arteries filled with blue latex and pulmonary veins filled with red latex?

3. Note that the cat has only two branches exiting from the aortic arch. The first branch is the brachiocephalic trunk. Follow the brachiocephalic trunk superiorly to the point where two branches exit the artery in a V-shape. These vessels are the right and left common carotid arteries.

   ▶ How is this configuration different than in the human?

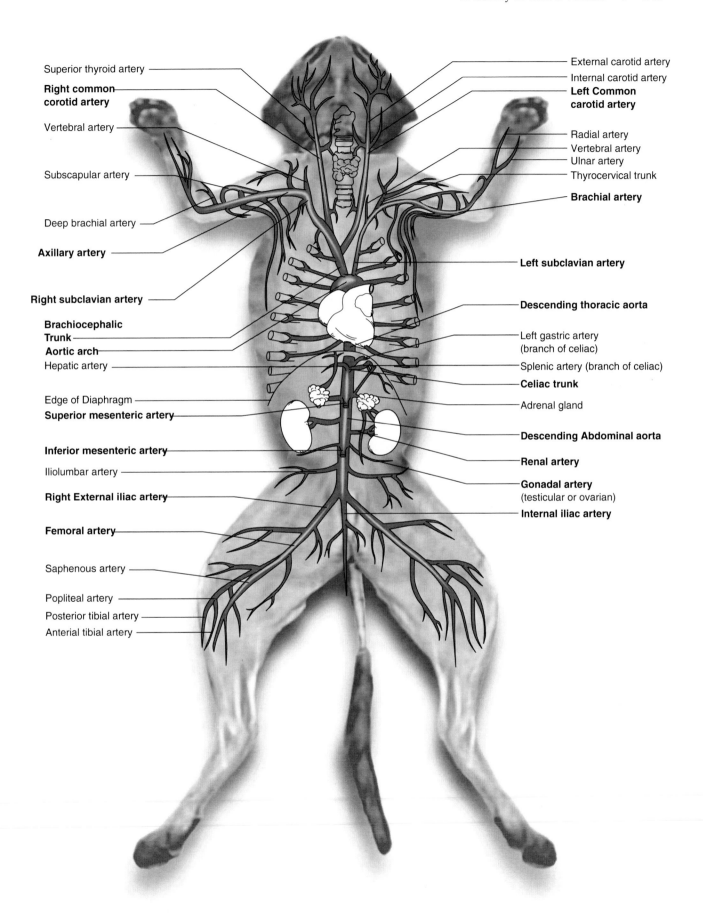

Superior thyroid artery

Right common corotid artery

Vertebral artery

Subscapular artery

Deep brachial artery

Axillary artery

Right subclavian artery

Brachiocephalic Trunk

Aortic arch

Hepatic artery

Edge of Diaphragm

Superior mesenteric artery

Inferior mesenteric artery

Iliolumbar artery

Right External iliac artery

Femoral artery

Saphenous artery

Popliteal artery

Posterior tibial artery

Anterial tibial artery

External carotid artery

Internal carotid artery

Left Common carotid artery

Radial artery

Vertebral artery

Ulnar artery

Thyrocervical trunk

Brachial artery

Left subclavian artery

Descending thoracic aorta

Left gastric artery (branch of celiac)

Splenic artery (branch of celiac)

Celiac trunk

Adrenal gland

Descending Abdominal aorta

Renal artery

Gonadal artery (testicular or ovarian)

Internal iliac artery

FIGURE 21.11 ◆ Cat Arteries

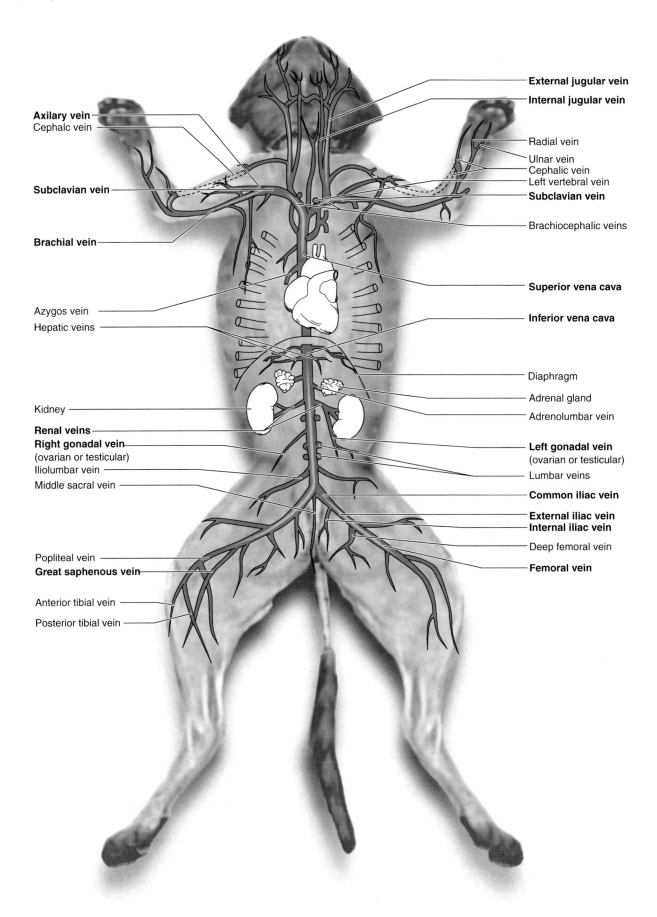

Axilary vein
Cephalc vein

Subclavian vein

Brachial vein

Azygos vein
Hepatic veins

Kidney

Renal veins
Right gonadal vein
(ovarian or testicular)
Iliolumbar vein
Middle sacral vein

Popliteal vein
Great saphenous vein

Anterior tibial vein
Posterior tibial vein

External jugular vein
Internal jugular vein

Radial vein
Ulnar vein
Cephalic vein
Left vertebral vein
Subclavian vein

Brachiocephalic veins

Superior vena cava

Inferior vena cava

Diaphragm
Adrenal gland
Adrenolumbar vein

Left gonadal vein
(ovarian or testicular)
Lumbar veins

Common iliac vein

External iliac vein
Internal iliac vein
Deep femoral vein

Femoral vein

FIGURE 21.12 ◆ Cat Veins

4. After the common carotid arteries exit the brachiocephalic trunk, the vessel changes its name to the right **subclavian** a. In the axillary region the vessel becomes the **axillary** a. The axillary a. continues into the brachium as the **brachial** a.

5. Follow the course of the descending aorta in the thorax where it passes through the diaphragm. You will need to cut the parietal peritoneum to see the aorta against the posterior abdominal wall. Trace the **abdominal aorta** and identify its branches.

6. The superior most branch exiting from the anterior surface of the abdominal aorta is the **celiac trunk.**

7. The next branch off the anterior surface of the abdominal aorta inferior to the celiac trunk is the **superior mesenteric** a. Follow the superior mesenteric a. to its attachment to the mesenteries of the small intestine.

8. The **renal arteries** exit the lateral abdominal aorta and lead to the kidneys.

9. A third unpaired artery exiting from the anterior surface of the abdominal aorta is the **inferior mesenteric** a. This artery often rips if the intestines are pulled too far out of the abdominal cavity. Lateral to the inferior mesenteric are the paired **gonadal arteries.**

10. Within the pelvic cavity the abdominal aorta splits into three branches: a right and left **external iliac** a. and a central unpaired **internal iliac** a.

    ▶ How is this different than the human iliac arteries?

11. When the external iliac a. passes through the body cavity wall to the thigh it becomes the **femoral** a.

Veins

1. The **femoral** v. lies next to the femoral a. Exiting the femoral v. at the knee is the **great saphenous** v., an obvious superficial vein positioned on the media surface of the leg.

2. Superiorly the femoral v. passes into the pelvic cavity to become the **external iliac** v. The **internal iliac** v. exits medially and posteriorly from the external iliac v. Superior to this branch the vessel becomes the **common iliac** v. The right and left common iliac veins converge to form the **inferior vena cava** deep to the V formed by the external iliac arteries.

3. Follow the inferior vena cava superiorly and identify the branches that join with it. The **renal veins** drain the kidneys. The **left gonadal** v. joins the left renal v. The **right gonadal** v. attaches to the inferior vena cava inferior to the right renal v.

4. Observe the tissue located between the duodenum of the small intestine and the liver. The **hepatic portal** v. is a subtle thin tan vessel that drains veins exiting digestive viscera, and delivers the nutrient rich blood to the liver.

    ▶ This vein will not have blue latex in it. Why?

5. The inferior vena cava continues superior to the liver to join the heart posteriorly.

6. The **superior vena cava** extends superiorly from the heart. It branches into a right and left **brachiocephalic** v. On both sides, the **external jugular vein** branches superiorly up the neck from the brachiocephalic v. A small diameter **internal jugular** v. exits medially and superiorly from the larger external jugular v.

7. Beyond the external jugular v., the brachiocephalic vessel becomes the **subclavian** v. In the axillary region the subclavian v. becomes the **axillary** v., which becomes the **brachial** v. in the brachial region.

III. Special Circulations

A. Pulmonary Circulation (Figure 21.13)

The pulmonary circulation differs in many ways from systemic circulation, because it does not serve the needs of the body tissues with which it is associated (in this case, lung tissue.) It functions instead to bring the blood into close contact with the alveoli of the lungs to permit gas exchange. (The blood supply to the lungs is provided by the bronchial arteries, which diverge from the thoracic portion of the descending aorta.)

Be able to trace the pathway of blood flow in the pulmonary circulation and to identify the following tributaries of the pulmonary circulation:

pulmonary trunk
Rt/Lt pulmonary arteries
lobar arteries
Rt/Lt pulmonary veins

B. Hepatic Portal System (Figure 21.14)

The portal system drains the gastrointestinal tract and delivers this blood to the liver for processing via the hepatic portal vein. If a meal has recently been eaten, the hepatic portal blood will be nutrient rich and may be toxicant rich. The liver is the key organ involved in maintaining proper nutrient concentration, detoxifying toxicants, and removing bacteria and other debris from the passing blood. The liver in turn is drained by the hepatic veins that enter the inferior vena cava.

Be able to trace the pathway of blood flow through the hepatic portal system and to identify the following tributaries of this system:

hepatic vein
hepatic portal vein
splenic vein
superior mesenteric vein
inferior mesenteric vein

C. Fetal Circulation

In a developing fetus the lungs and digestive system are not yet functional, and all nutrient, excretory, and gas exchanges occur through the placenta. Nutrients and O_2 move across placental barriers from the mother's blood into fetal blood, and CO_2 and other metabolic wastes move from the fetal blood to the mother's blood. Fetal blood travels through the umbilical cord, which contains 3 blood vessels: one **umbilical vein** (carries nutrients and O_2 to the fetus) and two **umbilical arteries** (carry wastes and CO_2 from the fetus to the placenta). These vessels meet at the **umbilicus** and form the **umbilical cord.**

Oxygenated blood flows through the umbilical vein superiorly to the inferior vena cava, and then to the right atrium. Because fetal lungs are nonfunctional and collapsed, two shunting mechanisms ensure that blood almost entirely bypasses the lungs. Much of the blood entering the right atrium is shunt into the left atrium through the **foramen ovale,** an opening in the interatrial septum. The left ventricle then pumps the blood out the aorta to the systemic circulation. Blood that does enter the right ventricle and is pumped out of the pulmonary trunk encounters a second shunt, the **ductus arteriosus,** a short vessel connecting the pulmonary trunk and the aorta. Because the collapsed lungs present an extremely high-resistance pathway, blood more readily enters the systemic circulation through the ductus arteriosus. The aorta carries blood to the tissues of the body; this blood ultimately finds its way back to the placenta via the umbilical arteries. At birth, or shortly after, the foramen ovale closes and becomes the **fossa ovalis,** and the ductus arteriosus collapses and becomes the **ligamentum arteriosum.** After birth, lack of blood flow through the umbilical vessels leads to their eventual obliteration, and the circulation pattern becomes that of the adult.

Be able to identify the following structures, and to trace the pathway of fetal blood flow in figures or lab charts:

umbilical vein
umbilical arteries
umbilical cord
umbilicus
foramen ovale
ductus arteriosus

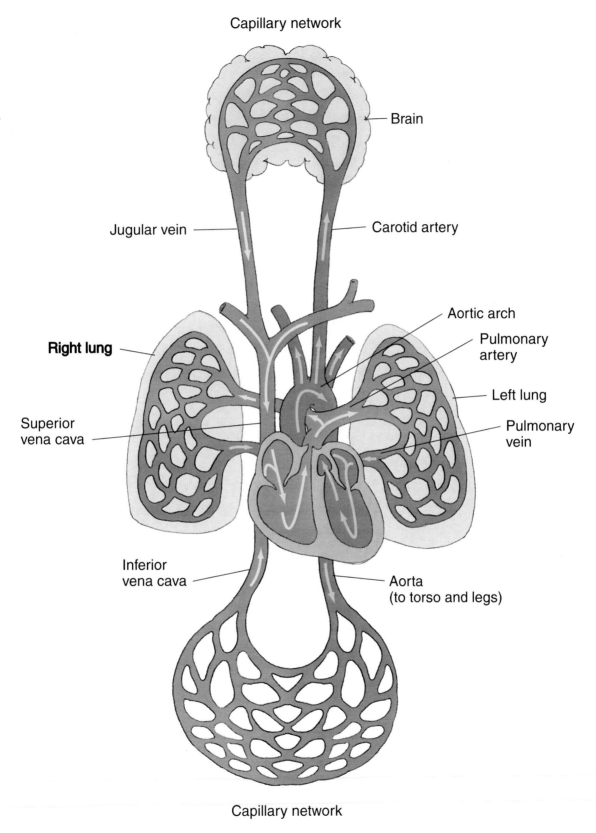

Capillary network

Brain

Jugular vein

Carotid artery

Aortic arch

Pulmonary artery

Right lung

Left lung

Superior vena cava

Pulmonary vein

Inferior vena cava

Aorta (to torso and legs)

Capillary network

FIGURE 21.13 ◆ Diagrammatic Circulation

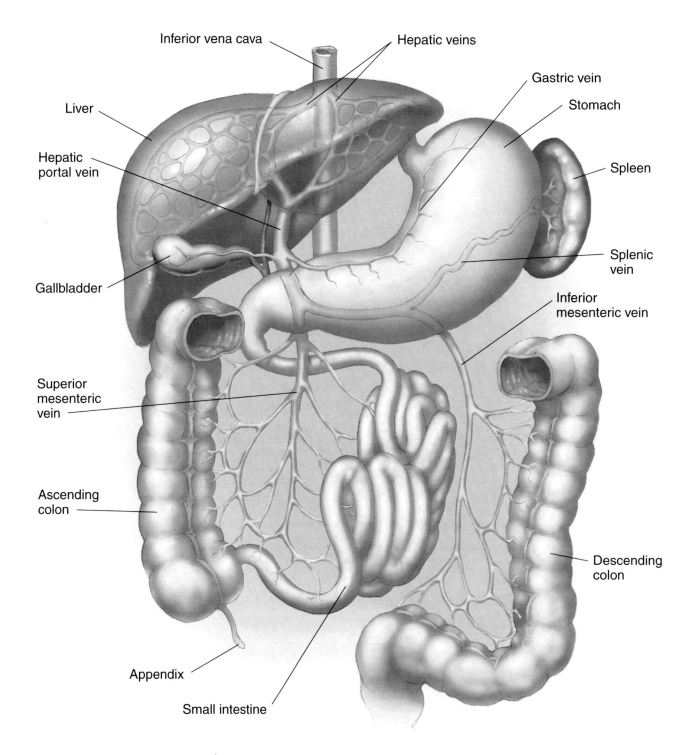

FIGURE 21.14 ◆ Hepatic Portal System

Name _____

Lab 21: Anatomy of Blood Vessels

◆ **Practice**

1. Describe the features of the three types of blood vessels in the table below.

Feature	Artery	Vein	Capillary
Tunica interna (intima)			
Tunica media			
Tunica externa (adventitia)			
Lumen size in adjacent arteries and veins			
Valves (yes or no)			
Function			

2. Choose the correct artery for each of the descriptions below:

    a. brachial
    b. celiac trunk
    c. dorsalis pedis
    d. fibular
    e. hepatic
    f. inferior mesenteric
    g. internal carotid
    h. left common carotid
    i. left subclavian
    j. popliteal
    k. posterior tibial
    l. pulmonary trunk
    m. superficial palmar arch
    n. superior mesenteric

_____ Supplies the distal large intestine

_____ Provides major blood supply for the brain

_____ Second branch off the aortic arch

_____ Transports oxygen poor blood away from heart

_____ Fed from the axillary artery

_____ Serves the knee region

_____ First branch off the abdominal aorta inferior to diaphragm

_____ Serves the lateral leg

_____ Serves the foot

_____ Serves the hand

3. Into which common vein does blood flow from the digestive organs?

4. The above vein delivers blood to which organ? For what purpose?

5. Trace the flow of blood through fetal circulation. What types of substances can cross the placenta?

6. What is the function of venous valves?

BLOOD PRESSURE

Lab 22

Objectives

1. Define and explain **systolic** and **diastolic** blood pressure
2. Describe the process of measuring blood pressure using the auscultatory method and accurately determine a subject's blood pressure using the auscultatory method
3. Relate the systolic and diastolic pressures to ventricular events in the heart
4. Investigate the changes of blood pressure in different conditions, and explain the functional significance of these changes

Materials

- Blood pressure cuffs
- Stethoscopes
- Alcohol swabs

INTRODUCTION

Blood pressure is the force exerted upon the walls of blood vessels by the fluid and cells within blood. As the ventricles of the heart contract, blood is forcefully ejected into the elastic arteries. This movement affects downstream blood and continues blood flow in the entire circulation. Because the ventricles rhythmically contract and relax, the blood vessels experience regular increases and decreases in pressure. **Systolic blood pressure** (SBP) is the highest pressure in the artery, and is produced as a result of ventricular contraction (systole). Normal values for healthy adults range from 110 to 140 mmHg. **Diastolic blood pressure** (DBP) is the lowest pressure in the artery, and occurs during ventricular relaxation (diastole). Healthy adults can experience diastolic pressures of 75 to 80 mmHg. In this lab you will practice measuring blood pressure and compare pressures of a subject at rest, after exercise and in various positions.

CONCEPTS

Measurement of Blood Pressure

Measurement of blood pressure provides useful information about heart and blood vessel condition. Generally, the SBP indicates the force and efficiency with which the heart is pumping blood. The DBP is an indication of the elasticity of blood vessels. A relatively easy and noninvasive method of measuring blood pressure is the **auscultatory method.**

The word *auscultate* means listen. The auscultatory method of measuring blood pressure involves correlating the sounds of blood flow within blood vessels to pressures measured on a pressure gauge. A **sphygmomanometer,** commonly called a **blood pressure cuff** (BP cuff), is used to stop blood flow through the brachial artery by increasing pressure superior to the elbow. A **stethoscope** amplifies sounds of blood flow through the brachial artery as pressure from the BP cuff is slowly released. Initial blood flow will be turbulent as blood begins to spurt through the constricted vessel. The audible sounds are called Kortokoff sounds, and begin when the BP cuff pressure is the same as the SBP. As pressure is released and the brachial artery fully opens, the blood flow settles to a smooth flow and the sounds disappear. The pressure on the BP cuff gauge when sounds disappear is the DBP.

◆ **Exercise 22.1** *Measure Blood Pressure Using the Auscultatory Method*

1. Work in pairs. Secure the BP cuff superior to the elbow of your partner and inflate the cuff to about 160mmHg. This pressure should exceed most healthy adults' SBP.
2. Place the bell of the stethoscope on the antecubital region where the brachial artery divides into the radial and ulnar arteries. You should hear no sounds.
3. Slowly release the pressure on the BP cuff while observing the pressure gauge.
4. Take note of the pressure on the gauge when you first begin to hear turbulent blood flow in the brachial artery. This pressure is the SBP.
5. Take note of the pressure on the gauge when you cease to hear any sounds in the stethoscope. This pressure is the DPB.
6. Swab the ear pieces of the stethoscope with alcohol when finished.

◆ **Exercise 22.2** *Measure the Effects of Position and Exercise on Blood Pressure*

1. In this exercise one lab partner will always measure BP and one partner will be the "patient" whose pressure will always be measured.
2. Measure the patient's blood pressure while the patient is sitting and record the pressure in the table below.
3. Measure the patient's blood pressure while the patient is lying down and record the pressure in the table below.
4. Measure the patient's blood pressure while the patient is standing and record the pressure in the table below.
5. Have the patient run in place for about two minutes. Measure the patient's blood pressure immediately after running. Record the pressure in the table below.
6. Swab the ear pieces with alcohol when finished.

	Sitting	Lying	Standing	After running
Blood Pressure (mmHg)				

▶ What are the effects of position and exercise on blood pressure?

▶ What are the advantages of altered blood pressure for different activities?

Name _____

Lab 22: Blood Pressure

◆ Practice

1. Define *blood pressure*.

2. What events of the cardiac cycle cause SBP and DBP?

3. What instruments are used in the auscultatory method of measuring blood pressure?

4. Explain the significance of the sounds heard while measuring blood pressure using the auscultatory method.

5. Hypertension is high blood pressure. Give several reasons why hypertension is dangerous to a person's health.

6. Hypotension is low blood pressure. Although a low blood pressure is common among healthy, athletic individuals, too low a blood pressure can be life threatening. Why?

ANATOMY OF THE RESPIRATORY SYSTEM

Objectives

1. Identify major respiratory structures on models, specimens and diagrams
2. Diagram the histology of the trachea in cross sectional diagrams
3. Explain the functions of major respiratory structures

Materials

- Lung model
- Torso model
- Cat specimens
- Dissecting trays
- Dissecting instruments
- Disposable gloves

INTRODUCTION

Most every cell in the body utilizes oxygen (O_2) in the metabolic processes that synthesize energy-storing ATP. (Red blood cells use anaerobic metabolism). This system requires a mechanism for oxygen delivery to all tissues, from superficial to deep. The respiratory system works in close association with the cardiovascular system to deliver oxygen to all tissues. During metabolism, cells produce carbon dioxide (CO_2) as a waste product that must be eliminated from the body. Again, the cardiovascular system coordinates with the respiratory system to remove CO_2 from the tissues and transport it to the lungs where it can be expired into the atmosphere.

Structures in the respiratory system assist with two major processes: ventilation and respiration. **Ventilation** is the movement of air into and out of the lungs. It is accomplished by the contraction and relaxation of muscles that affect the size of the thoracic cavity, and consequently the size of the lungs. **Respiration** is the movement of gas molecules across membranes. External respiration involves the diffusion of O_2 molecules from the lungs into the pulmonary capillaries and of CO_2 molecules from the pulmonary capillaries into the lungs. Internal respiration occurs at the body's tissues. O_2 diffuses from the systemic capillaries into the tissues and CO_2 moves from the tissues into the systemic capillaries. In this lab we will examine structures within the respiratory system and review their functions.

CONCEPTS

I. Respiratory System Structures

A. Survey of Respiratory Structures

Respiratory system structures are either part of the conducting zone or the respiratory zone. The **conducting zone** includes structures from the nasal cavity to the terminal bronchioles. These structures serve as air passageways, and also warm and humidify the air. **Respiratory zone** structures begin at the respiratory bronchioles and continue to the respiratory membrane. This is the site of gas exchange. Specific structures and their functions are listed in Table 23.1.

Table 23.1
Functions of Respiratory Structures

◆◆◆

Structure	Features	Function
Nostril	Opening from the nasal cavity to the atmosphere	Allows passage of air into and out of the nasal cavity
Nasal conchae	Three ridges on the lateral aspects of the nasal cavity covered with nasal mucosa	Creates turbulence in inspired air to increase contact with mucosa
Posterior nasal aperture	Opening between nasal cavity and nasopharynx	Allows passage of air between nasal cavity and nasopharynx
Uvula	Portion of soft palate that extends into the oropharynx	Closes the passageway between the nasopharynx and oropharynx upon swallowing
Nasopharynx	Air passageway posterior to the nasal cavity	Allows air to pass between the nasal cavity and the oropharynx
Pharyngeal tonsils	Lymphoid tissue within the lateral walls of the nasopharynx	Traps pathogens inhaled through the nasal cavity and houses immune cells that mount an immune response against the pathogens
Oropharynx	Passageway posterior to the oral cavity	Allows air, fluid and food to pass from the oral cavity to the laryngopharynx
Palatine tonsils	Lymphoid tissue within the lateral walls of the oropharynx	Traps pathogens and houses immune cells to mount an immune response against them
Lingual tonsil	Lymphoid tissue on the posterior tongue in the oropharynx	Traps pathogens and houses immune cells to mount an immune response against them
Laryngopharynx	Passageway inferior to the oropharynx; continuous with the esophagus	Passageway for air in and out of the larynx, and food into the esophagus
Larynx	Passageway bound by cartilages between the laryngopharynx and larynx	Contains the **epiglottis** that directs air and food or fluid into the proper passageway; contains the **vocal folds** that vibrate and produce a voice

B. **Larynx (Figure 23.2)**

The **larynx** is commonly called the voicebox. It is comprised of several cartilages joined by ligaments. The superior border is defined by the **hyoid bone,** the only bone in the body that does not articulate with another bone. It does articulate with the thyroid cartilage inferiorly, however. The **thyroid cartilage** is broader anteriorly with a V-shaped superior ridge. An anterior protrusion, the **laryngeal prominence,** forms the "Adam's apple." Inferior to the thyroid cartilage is the butterfly shaped **cricoid cartilage.** Posteriorly the **epiglottis** is poised superior to the larynx and closes the opening of the larynx during swallowing. The epiglottis is formed from elastic cartilage. Two **arytenoid cartilages** in the posterior wall of the larynx act like tuning forks on a guitar to tighten or relax the **arytenoid muscles** attached to the **vocal folds.** In this way the vibrations of the vocal folds create sound as air passes over them, much like the strings of a guitar. The superiorly placed vestibular folds do not contribute to sound production. The inner surface of the larynx is covered with a mucus membrane of ciliated epithelium.

C. **Trachea (Figures 23.3 and 23.4)**

The **trachea** is formed from stacked C-shaped rings of **hyaline cartilage.** Posteriorly the gap in the Cs is closed by a long **trachealis muscle.** This muscle can expand into the trachea to allow passage of material in the posteriorly adjacent esophagus. The internal surface of the trachea is lined by a **mucosa** deep to which is a **submucosa.**

Structure	Features	Function
Trachea	Passageway for air between the larynx and primary bronchi; lined with ciliated mucus membrane; **C**-shaped rings of hyaline cartilage are bound posteriorly by the **trachealis muscle**	Allows passage of air between larynx and primary bronchi; cilia move mucus superiorly toward pharynx; trachealis muscle allows passage of food in adjacent esophagus
Primary bronchi	Bifurcation of the trachea; each bronchus supplies one lung; cartilage and smooth muscle	Distribute air to both lungs
Secondary bronchi	Branches from the primary bronchi; three on right and two on left; cartilage and smooth muscle	Supply air to each lobe of the lungs
Tertiary bronchi	Increasingly smaller branches; less cartilage and more smooth muscle	Distribute air to all regions of lungs
Bronchioles	Branches with diameter less than 1 mm; smooth muscle present but no cartilage	Constrict and dilate to control air movement to and from lungs
Terminal bronchiole	Passageway between bronchioles and respiratory bronchioles	Air passageway
Respiratory bronchioles	Tiniest bronchioles with alveolar air sacs attached	Air passageway into and out of alveoli and alveolar ducts
Alveolar duct	Passageway from which alveoli outpocket; attached to respiratory bronchioles	Air passageway into and out of alveoli
Alveolus	Outpocketing of alveolar duct; wall is one layer of epithelial cells; covered by pulmonary capillaries; capillary wall plus alveolus wall is the **respiratory membrane**	Gases diffuse across respiratory membrane; oxygen moves from alveolus into capillary and carbon dioxide moves from capillary into alveolus
Alveolar sac	Group of alveoli attached to the alveolar duct; analogous to a bunch of grapes	Site of gas exchange

External to the submucosa is the hyaline cartilage. The outside surface of the trachea is covered by a connective tissue **adventitia.**

D. **Lungs (Figure 23.5)**

The **lungs** are served by the **primary bronchi,** which branch into **secondary bronchi.** The right lung has three lobes while the left lung only has two lobes so as to accommodate the heart. Each lobe of the lung is served by one secondary bronchus. Primary bronchi and blood supply enter the lungs at an indentation called the **hilus.** The external surface of the lungs is covered by the **visceral pleurae.** The **parietal pleurae** line the inside of the thoracic cavity. The **pleural cavity** in between the parietal and visceral layers is filled with **pleural fluid.** It is the interaction between the pleurae, the pleural fluid and the lungs that keeps the lungs inflated.

E. **Respiratory Zone Structures (Figure 23.6)**

Gas exchange occurs across the microscopic respiratory membrane in alveolar sacs. Pulmonary capillaries surround the alveoli. The epithelial cells of the alveoli, the endothelial cells of the capillaries, and the intervening basement membrane together make up the respiratory membrane. There are more than 300 million alveoli that form the macroscopic structure of the lungs.

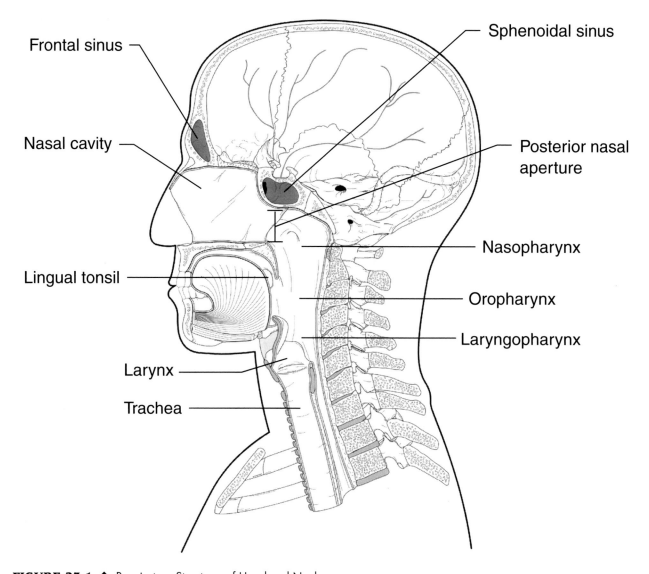

FIGURE 23.1 ◆ Respiratory Structures of Head and Neck

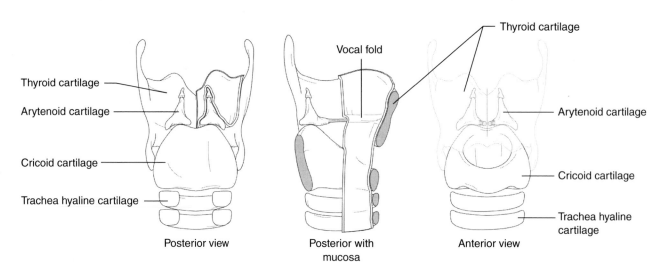

FIGURE 23.2 ◆ Larynx

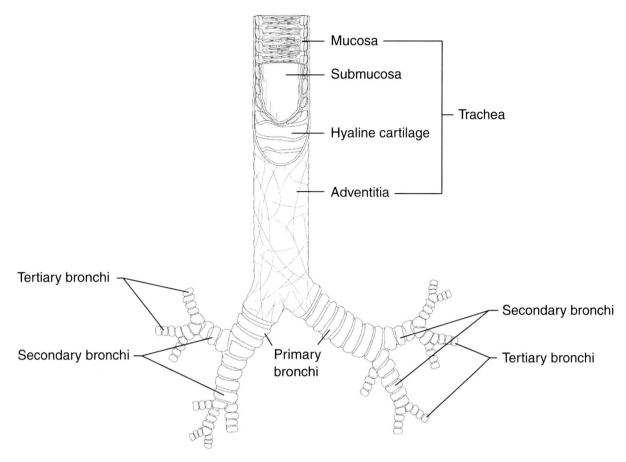

FIGURE 23.3 ◆ Trachea and Primary Bronchi

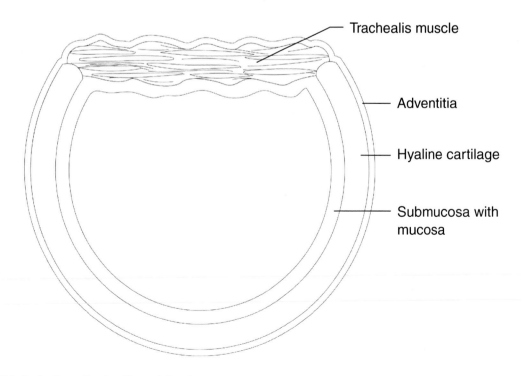

FIGURE 23.4 ◆ Cross Section Through Trachea

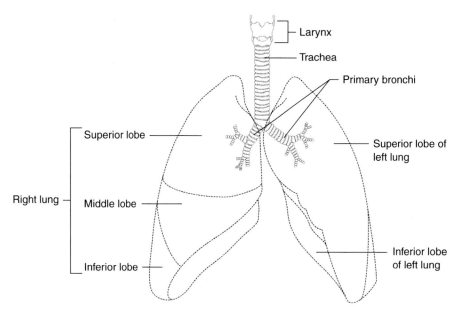

FIGURE 23.5 ◆ Lungs

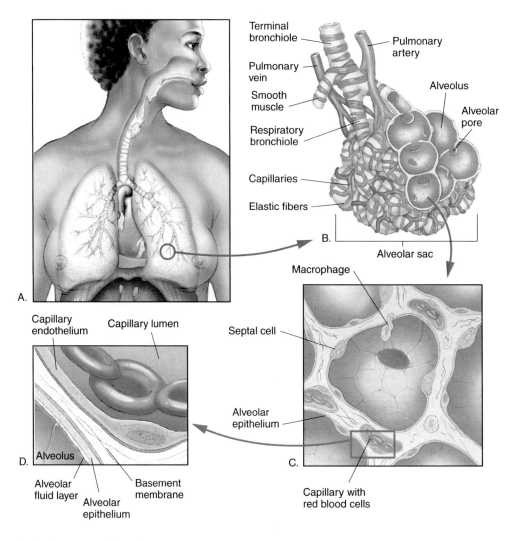

FIGURE 23.6 ◆ Respiratory Zone Structures

◆ **Exercise 23.1** *Identify Respiratory Structures on Human Lung, Larynx, and Torso Models*

Be able to identify the following structures in models of the human lung, larynx, and respiratory system (Photos 386a and 386b):

Nasal cavity	Arytenoid cartilage
Nostrils	Arytenoid muscle
Nasal conchae	Vocal folds
Posterior nasal aperture	Vestibular folds
Nasopharynx	Trachea
Oropharynx	Trachealis muscle
Laryngopharynx	Primary bronchi
Larynx	Secondary bronchi
Epiglottis	Lobes of the lungs
Thyroid cartilage	Hilus
Cricoid cartilage	

◆◆◆

◆ **Exercise 23.2** *Identify Respiratory Structures on a Cat Specimen*

Use Photo 411 as a guide for identification and dissection.

1. Obtain your preserved cat specimen and position it on a dissecting tray ventral side up.
2. Expose the laryngeal region in the ventral region of the neck. Identify the **thyroid gland** anterior and lateral to the **thyroid cartilage.** Immediately inferior to the thyroid cartilage is the **cricoid cartilage.** Make a longitudinal incision in the ventral aspect of the larynx and identify the **true vocal cords.**
3. Inferior to the larynx identify the **trachea.** Follow the trachea inferiorly to its bifurcation and identify the **primary bronchi.**
4. Multilobed **lungs** are positioned on both sides of the heart. Each lung is contained within a pleural cavity.
   ▶ How many lobes are in the right and left lungs of humans?

5. Note the diaphragm muscle that divides the thoracic and abdominal cavities.

◆◆◆

◆ **Exercise 23.3** *Identify Respiratory Structures in a Sheep Pluck, Larynx, and Trachea*

Use Photos 387 and 388 as a guide for identification.

1. A sheep pluck is a connected combination of the larynx, trachea, bronchi and lungs together with the heart. Often the esophagus and parts of the diaphragm muscle are attached. Observe the relationship of structures present in the sheep pluck. Identify as many structures as you can.

2. Separate sheep larynxes with tracheas are available for dissection. Examine the intact preparation and identify the following structures: **hyoid bone, epiglottis, thyroid cartilage, cricoid cartilage** (you may need to scrape away the adventitia to observe the cartilages), **trachea rings, trachealis muscle.**

3. Hemisect the larynx and trachea and identify the following structures: (you may need to scrape away the mucosa to observe internal structures) **arytenoid cartilage, arytenoid muscles, vocal fold, vestibular fold.**

◆◆◆

Name _____

Lab 23: Anatomy of the Respiratory System

◆ Practice

1. Draw from memory a cross section through the trachea. Label the histological layers.

2. The uvula and epiglottis have similar functions. Compare and describe the functions of both structures.

3. Why is it advantageous for the nasal conchae to cause turbulence in inspired air?

4. Explain the difference in function between conducting zone structures and respiratory zone structures.

5. What cell types are part of the respiratory membrane?

6. Which gases move through the respiratory membrane and in which direction?

7. By what mechanism does mucus move in respiratory structures? What is the purpose of the movement of mucus?

RESPIRATORY PHYSIOLOGY

Lab
24

Objectives

1. Define **respiratory volume** and **respiratory capacity**
2. Describe specific respiratory volumes and capacities
3. Use a wet spirometer to measure respiratory volumes and capacities
4. Calculate specific respiratory capacities
5. Describe the effects of gender and smoking on respiratory capacities

Materials

- Wet spirometers
- Cardboard mouth pieces

INTRODUCTION

Gas exchange is dependent upon adequate ventilation. The ability of the lungs to fill with air and to expire air dramatically affects respiration. Various conditions severely inhibit adequate ventilation and may require lifestyle changes or limitations. For example, lung infections may cause pneumonia. Pneumonia is a condition in which the lungs fill with fluid and decrease the functional surface area of the respiratory membrane. Another lung condition is emphysema in which the elastic fibers of the lungs are replaced with dense fibrous tissue that inhibits lung expansion, and in which the alveoli are gradually destroyed. Patients with emphysema feel like they can't "catch their breath" and eventually have to carry a supplemental oxygen tank with them. The primary cause of emphysema is years of exposure to tobacco smoke.

Efficiency of lung function can be measured through evaluation of four main **respiratory volumes.** Different combinations of respiratory volumes describe specific **respiratory capacities.** In this lab we will use a simple but effective instrument called a **wet spirometer** to measure different respiratory volumes and capacities, and use this data to calculate others.

Table 24.1
Respiratory Volumes

Respiratory Volume	Description	Adult Male Average Value	Adult Female Average Value
Tidal Volume (TV)	The amount of air that moves into (then out of) the lungs in a normal quiet breath	500 ml	500 ml
Inspiratory Reserve Volume (IRV)	The amount of air that can be forcibly inhaled after a tidal volume inhalation	3100 ml	1900 ml
Expiratory Reserve Volume (ERV)	The amount of air that can be forcibly exhaled after a tidal volume exhalation	1200 ml	700 ml
Residual Volume (RV)	Amount of air remaining in the lungs after a forceful exhalation; prevents alveolar collapse	1200 ml	1100 ml

271

CONCEPTS

The total lung capacity is the sum of four respiratory volumes.
The respiratory capacities are combinations of the respiratory volumes.

Table 24.2
Respiratory Capacities

Respiratory Capacity	Description	Adult Male Average Value	Adult Female Average Value
Total Lung Capacity (TLC)	Total amount of air the lungs can possibly hold TLC = TV + IRV + ERV + RV	6000 ml	4200 ml
Vital Capacity (VC)	The total amount of air that can be exhaled after maximum inhalation VC = TV + IRV + ERV	4800 ml	3100 ml
Inspiratory Capacity (IC)	The total amount of air that can be inhaled after a tidal volume exhale IC = TV + IRV	3600 ml	2400 ml
Functional Residual Capacity (FRC)	The amount of air remaining in the lungs after a tidal volume exhale FRC = ERV + RV	2400 ml	1800 ml

◆ Exercise 24.1 *Measure Respiratory Volumes*

Work in groups for this exercise. One person will need to record the data and another will act as the subject. Every person in the group should take turns to use the spirometer to collect data about respiratory volumes and record the data in the class chart.

1. Without using the spirometer, count and record the subject's normal respiratory rate.

   Respirations per minute _____

2. Obtain a disposable cardboard mouthpiece. Insert it in the open end of the flexible tube of the wet spirometer. Before beginning, the subject should practice exhaling through the mouthpiece without exhaling through the nose.

3. Conduct each test three times for each measurement. Remember to reset the spirometer indicator to zero before beginning each trial. Record the data in the space below, and then calculate the average volume for that respiratory measurement. After you have completed the trials and calculated the averages, enter the average values for VC on the table prepared on the whiteboard for tabulation of class data. You will enter your value on the table twice; once for your gender and once for your smoking status. Transfer the entire class data to the table on page 274 and calculate the class averages.

4. To conduct the test for TV, inhale a normal breath, and then exhale a normal breath of air into the spirometer mouthpiece. (Do not force the expiration!) Record the volume and repeat the test twice.

   Trial 1 _____ml

   Trial 2 _____ml

   Trial 3 _____ml

   Average TV _____ml

5. ERV can be measured as follows. Inhale and exhale normally two or three times. Now exhale normally and before inhaling insert the spirometer mouthpiece and exhale forcibly as much of the additional air as you can. Record your results, and repeat the test twice again.

   Trial 1 _____ml

   Trial 2 _____ml

   Trial 3 _____ml

   Average ERV _____ml

6. VC can be measured by performing this test. Breathe in and out normally two or three times, and then bend over and exhale all the air possible. Then, as you raise yourself to the upright position, inhale as fully as possible. It is very important to strain to inhale the maximum amount of air that you can. Quickly insert the mouthpiece, and exhale as forcibly as you can. Record your results and repeat the test twice again.

   Trial 1 _____ml

   Trial 2 _____ml

   Trail 3 _____ml

   Average VC _____ml

7. IRV can now be calculated using the average values obtained for TV, ERV, and VC by plugging them into the equation: IRV = VC − (TV + ERV)

   Record your average IRV: _____ ml

   How does your calculated value compare to average values in Table 24.1?

Collect class data in the following table. Write the **VC** for each individual under the appropriate columns.

Males	Females	Smokers	Nonsmokers
Average:	Average:	Average:	Average:

Lab 24: Respiration Physiology

◆ Practice

1. Define the following terms:

   Tidal Volume

   Inspiratory Reserve Volume

   Expiratory Reserve Volume

   Residual Volume

2. What is the difference between a respiratory volume and a respiratory capacity?

3. Write the equation for each of the respiratory capacities listed below:

   Total lung capacity

   Vital capacity

   Inspiratory capacity

   Functional residual capacity

4. Compare the average VC values of smokers to nonsmokers. How many smoking and nonsmoking subjects are there in the class? Is this enough to calculate reliable results?

5. What do you think happens to respiratory volumes when you have a chest cold? Why?

6. What is the relationship between pulmonary ventilation and respiration?

ANATOMY OF THE DIGESTIVE SYSTEM

Lab 25

Objectives

1. Identify the major digestive organs and accessory structures on diagrams, models, and preserved cats
2. Describe the general functions of digestive system structures
3. Identify the general histologic layers of the wall of the alimentary canal
4. Identify specific microscopic structures in digestive organs

Materials

- Human torso model
- Digestive system model
- Microscope slides
  - Esophagus
  - Stomach
  - Small intestine
  - Large intestine
  - Liver
- Cat specimens
- Dissecting trays
- Dissecting instruments
- Disposable gloves

INTRODUCTION

The digestive system enables us to ingest foods, break them down into absorbable units, absorb the nutrients and eliminate the indigestible substances. Structures within the digestive system can be classified into two main categories: digestive tract organs and accessory organs. The digestive tract is also called the **gastrointestinal (GI) tract** and the **alimentary canal.** The GI tract begins at the oral cavity and continues as a muscular tube that ends at the anus. As food materials pass through the GI tract they are acted upon by muscles and chemicals. **Accessory organs** are those structures through which food materials do not pass, but which contribute to the digestive processes occurring in the GI tract.

Several processes contribute to digestion. **Ingestion** is the intake of food and drink. **Propulsion** can take the form of swallowing and **peristalsis,** which is the rhythmic contraction of smooth muscle to propel food materials distally in the GI tract. **Mechanical digestion** is the physical breakdown of food materials into absorbable units. Chewing and muscular **segmentation** movements contribute to mechanical digestion. **Absorption** occurs mostly in the small intestine, and involves movement of nutrients and other substances into the blood stream or lymph. Finally, **defecation** is the elimination of indigestible substances from the GI tract.

CONCEPTS

I. Major Organs of the GI Tract

Table 25.1 lists the features and functions of the major organs of the GI tract.

Table 25.1
Organs of the GI Tract

◆◆◆

GI Tract Organ	Features	Functions
Oral cavity	Site of ingestion; contains **teeth, tongue** and **salivary glands**	Teeth chew and break down food, salivary glands begin chemical digestion of carbohydrates; swallowing propels food bolus into the pharynx
Oropharynx	Region of pharynx posterior to oral cavity	**Uvula** prevents superior movement during swallowing, directs food bolus into laryngopharynx
Laryngopharynx	Region of pharynx superior to larynx and esophagus	**Epiglottis** of larynx closes during swallowing to direct food bolus into esophagus
Esophagus	Transport tube for food between pharynx and stomach; layers are inner mucosa and deeper submucosa, smooth muscle and outer adventitia; joins the stomach at the **cardiac sphincter (gastroesophageal sphincter)**	Mucus reduces friction as food bolus moves inferiorly, peristalsis assists movement of bolus
Stomach	"Way station" for food substances, about 8 in long, can hold about 1 gallon; inner mucosa is folded into **rugae, gastric pits** within the mucosa contain enzyme secreting cells; muscular region has three layers: **longitudinal, circular** and **oblique;** regions of stomach are rounded top called **fundus, body, pyloric canal** and **pylorus**	Food bolus is converted into white paste called **chyme;** protein digestion begins by digestive enzymes, segmentation mechanically breaks down food; rugae allow stomach to expand its volume
Small intestine	Joins the stomach at the **pyloric sphincter;** first section is the **duodenum,** about 10 in long; middle section is the **jejunum,** is about 8 ft long; third section is the **ileum,** is about 12 ft long; inner mucosa has circular folds called **plicae circulares;** finger like **villi** extend from mucosa into lumen to increase absorptive surface area; capillaries and lymph vessels called **lacteals** absorb substances through epithelial lining; **intestinal crypts** in mucosa contain enzyme secreting cells	Most digestion occurs here; enzymes from the pancreas continue protein and carbohydrate digestion; plicae circulares swirl chyme to increase contact with absorptive surface; nutrients and other substances absorbed into blood capillaries; bile from liver emulsifies fats and fatty substances are absorbed into lacteals
Large intestine	**Cecum** connects to ileum at **ileocecal valve; appendix** is attached to cecum; **ascending colon** rises to liver and turns medially at the **hepatic flexure** to continue as **transverse colon** inferior to stomach; **splenic flexure** turns into **descending colon; sigmoid colon** is S-shaped curve that ends in **rectum** and **anus;** external longitudinal bands of smooth muscle are **teniae coli,** which pull walls of large intestine into pockets called **haustra;** no plicae circulares or villi are present	Main functions are water absorption and transport to anus for elimination of indigestible residues; bacteria are housed here, called bacterial flora, which digest some carbohydrate residues and produce acids and gases; appendix at cecum houses immune cells but can easily become infected itself; presence of feces in rectum will stimulate defecation reflex
Anus	Short canal through the body cavity wall; lined with mucus membrane	Secretion of mucus eases passage of feces

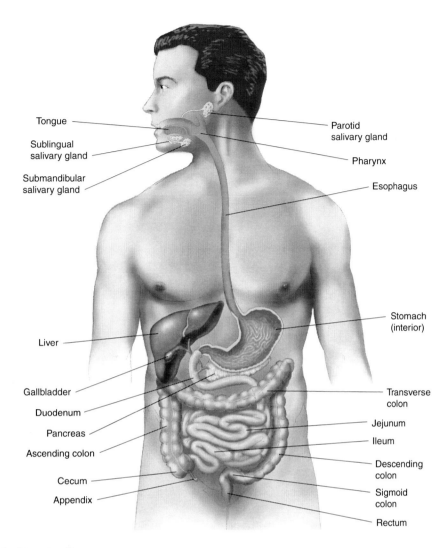

FIGURE 25.1 ◆ Digestive System

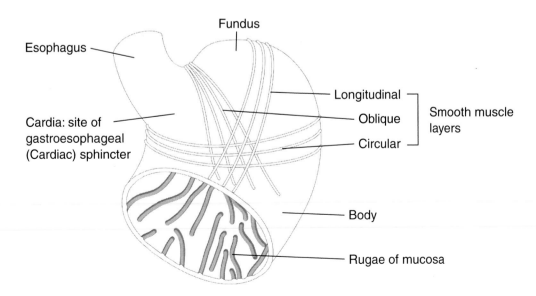

FIGURE 25.2 ◆ Transected Stomach

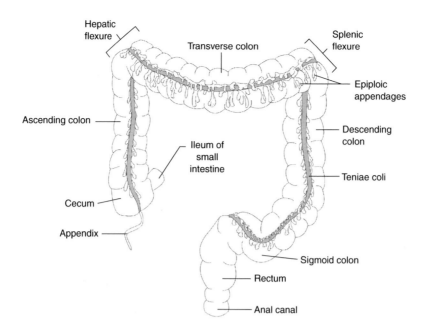

FIGURE 25.3 ◆ Large Intestine

II. Accessory Organs to the GI Tract

Table 25.2
Accessory Organs to the GI Tract

Accessory Organ	Location and Features	Functions
Teeth	In oral cavity at alveolar margin of maxillae and mandible	Chewing (mastication) for mechanical digestion to create a food bolus
Salivary glands	Extrinsic salivary glands include **parotid, submandibular, sublingual glands;** intrinsic glands are scattered within oral mucosa	Secrete saliva which contains **lysozyme** to inhibit bacteria and **amylase** to begin carbohydrate digestion; moistens food bolus; suspends taste particles
Tongue	Muscular structure in oral cavity; moved by extrinsic muscles	Manipulates food bolus to mix with saliva
Liver	Four lobes are: right, left, caudate and quadrate; manufactures **bile;** receives nutrient and toxicant rich blood from **hepatic portal vein; hepatic ducts** join to form **common hepatic duct**	Bile emulsifies fat for easier absorption; glucose removed from blood and stored as glycogen; toxins removed from blood and detoxified
Gall bladder	Nestled in ventral aspect of liver on right side; **cystic duct** joins common hepatic duct to form **bile duct** which ends at duodenum as the **hepatopancreatic ampulla and sphincter**	Stores bile for use as needed
Pancreas	Tadpole shaped structure inferior to stomach on left side; contains cells organized in pancreatic islets that secrete digestive enzymes; **main pancreatic duct** joins hepatopancreatic ampulla; **accessory pancreatic duct** joins duodenum directly	Digestive enzymes are activated in small intestine and chemically break down proteins and carbohydrates

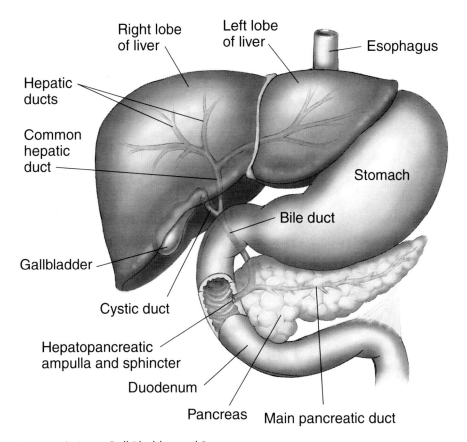

Right lobe of liver

Left lobe of liver

Esophagus

Hepatic ducts

Common hepatic duct

Stomach

Bile duct

Gallbladder

Cystic duct

Hepatopancreatic ampulla and sphincter

Duodenum

Pancreas   Main pancreatic duct

FIGURE 25.4 ◆ Liver, Gall Bladder, and Pancreas

◆ **Exercise 25.1** *Identify GI Tract Organs and Accessory Organs in Models*

Review the anatomy of the digestive system on a torso model and digestive tract model. Identify all bold faced structures in Tables 25.1 and 25.2 and be able to describe their functions.

◆◆◆

III. Mesenteries and Omenta

The serous membrane of the abdominal cavity is called the **peritoneum.** The **parietal peritoneum** lines the inner surfaces of the abdominal cavity, however some organs lie outside of the peritoneum. These **retroperitoneal organs** include the kidneys, part of the pancreas, part of the large intestine, urinary bladder and reproductive organs. Intraperitoneal organs are covered by **visceral peritoneum.** The peritoneal cavity is the space between the layers of the peritoneum and is filled with peritoneal fluid.

In some regions the peritoneum becomes a double walled **mesentery** to help secure organs in place. One mesentery suspends the jejunum and ileum of the small intestine from the posterior abdominal wall and maintains the folded conformation of the small intestine. Blood and lymph vessels travel through the mesentery to reach the intestines. Extending from the lesser and greater curvatures of the stomach are specialized mesenteries called the

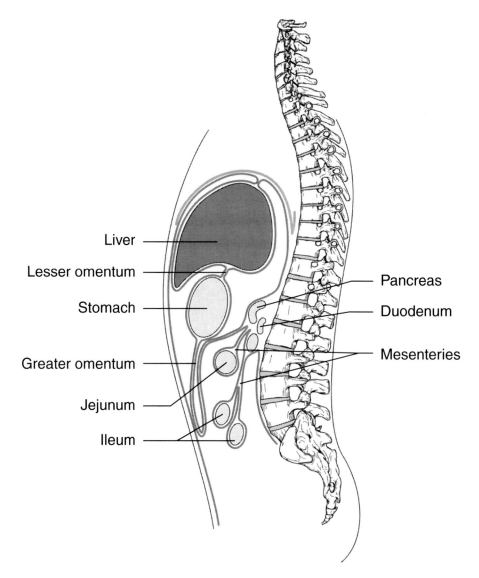

FIGURE 25.5 ◆ Mesenteries and Omenta

lesser and greater omenta respectively. The **lesser omentum** suspends the stomach from the liver while the **greater omentum** covers the coils of the small intestine. The greater omentum is a site for storage of abdominal fat.

◆ Exercise 25.2 *Identify Digestive Organs in a Cat Specimen*

1. Use Photos 409–410b and 412 to assist you in identifying digestive structures. In the neck region, posterior to the white trachea, find the muscular **esophagus.** Note its close association with the trachea.
2. Follow the esophagus inferiorly, posterior to the heart and through the **hiatus** in the diaphragm. This is the potential site of a hiatial hernia, in which the stomach is pulled upward through the hiatus into the thoracic cavity. The esophagus meets the **stomach** just inferior to the **liver.**
3. Gently examine the junction between the esophagus and the stomach. You will be able to feel the circular **gastroesophageal sphincter.** During your examination of the stomach you should identify the different regions, including the **cardia, fundus, body, pyloris,** and the **greater and lesser curvatures.** Cut open the stomach from the fundus to the pyloris and remove any contents to the trash. Now you should be able to see the mucosal folds of **rugae.**

4. Examine the **greater omentum** and its attachment to the greater curvature of the stomach. The greater omentum is a specialized region of serosa that contains fatty deposits. It covers the small intestines. Gently lift the omentum and fold it next to the stomach. You may have to lift the spleen away from the omentum to move it. Do not remove the omentum or the spleen.

5. At the pylorus of the stomach you will find the junction with the **duodenum** of the **small intestine.** Feel the tough, circular **pyloric sphincter** where the two organs meet. The duodenum receives chyme from the stomach as well as digestive juices from the liver and the pancreas. Using a blunt probe and your fingers, carefully dissect out, but do not cut, the **common bile duct** where it joins the duodenum.

6. Emerging from the duodenum with the common bile duct is the shorter **pancreatic duct.** Follow this duct inferiorly to the granular appearing **pancreas.**

7. Follow the short duodenum distally to the much longer **jejunum.** You will notice a difference in muscular thickness of the small intestine where the jejunum is continuous with the **ileum.** The ileum appears more wrinkled, the jejunum more smooth. Notice the membranous **mesentery** that holds the small intestines in place. The mesentery provides a route for blood and lymph vessels and for nerve supply.

8. Gently push the small intestine to the cat's right side to expose the large intestine. You will be able to identify the **ileocecal valve** at the junction between the ileum and **cecum** by feeling a thick knot of smooth muscle. The cecum is a large, pouch-like sac at the beginning of the large intestine. The large intestine in the cat does not have a clearly defined ascending colon or transverse colon. The descending colon ends in the **sigmoid colon,** which joins the **rectum.** The rectum then empties into the **anal canal.**

9. Examine the liver closely. How many lobes does the cat liver have? How many lobes does the human liver have? Tucked between two lobes on the right side you will find the thin-walled **gall bladder,** which may be green or tan. This sac stores bile, made by the liver, until it is released to the common bile duct via the **cystic duct.** Trace the cystic duct along its course to the common bile duct.

10. The **hepatic portal system** delivers nutrient and toxicant rich blood from the digestive viscera to the liver. Recall that a portal system delivers blood from an artery to a capillary bed to a vein, and then to another capillary bed. Because the **hepatic portal vein** is located between two capillary beds it will not be filled with blue latex. Usually it is a dark brown in color. Find the hepatic portal vein next to the common bile duct between the duodenum and the liver. The hepatic portal vein is fed by veins exiting digestive viscera. The liver removes many toxins and nutrients from the blood before sending it on to the inferior vena cava through the hepatic veins.

◆◆◆

IV. GI Tract Histologic Layers

The hollow organs of the GI tract, from esophagus to anal canal, have a similar histologic organization. In general, the lumen of the organs is lined by a **mucosa** that is a mucus secreting epithelial membrane. Digestive enzymes and hormones are also secreted from this layer. A highly vascular **submucosa** surrounds and supplies the mucosa. This connective tissue layer is rich with lymph vessels and nerve fibers. The **muscularis externa** is positioned deep to the mucosa. **Longitudinal** and **circular smooth muscle layers** create peristalsis and segmentation movements in the GI tract organs. Parasympathetic nerve supply maintains muscular activity while sympathetic supply exerts an inhibitory effect. Note that the stomach has a third smooth muscle layer in its muscularis externa: the oblique layer. With the exception of the esophagus, the external layer of the GI tract is the serous membrane, the **visceral peritoneum.** The esophagus, which is not located in the abdominal cavity, has a dense **adventitia** on the outer surface.

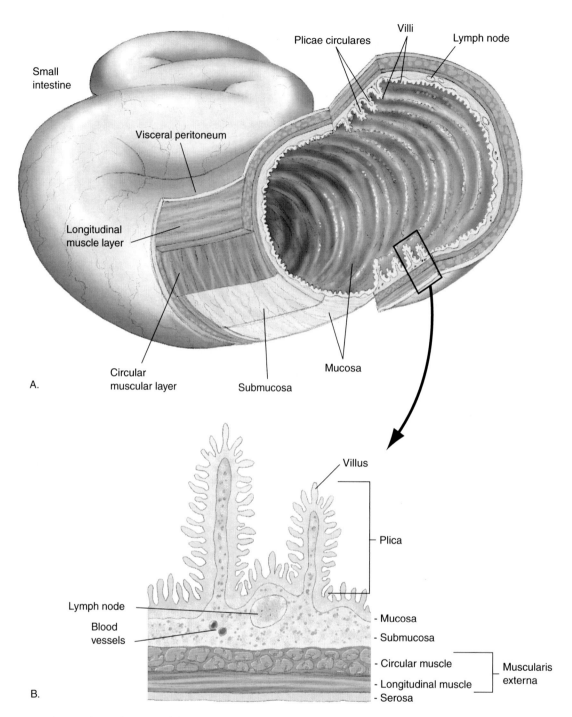

A.

B.

FIGURE 25.6 ◆ Histological Layers of the Small Intestine

◆ **Exercise 25.3** *Examine Microscopic Anatomy of Digestive Organs*

Esophagus

Examine a slide of esophagus tissue. Identify the following features and sketch and label what you see in the space below (Photo 389b):

Mucosa (stratified squamous epithelium)
Submucosa (areolar connective tissue, contains mucous glands)
Muscularis externa (circular layer, longitudinal layer)
Adventitia (fibrous connective tissue) (note that there is no serosa on the esophagus)

Stomach

Identify the following features and sketch and label what you see in the space below (Photos 390a and 390b):

Mucosa (simple columnar epithelium, gastric pits, gastric glands)
Submucosa (areolar connective tissue)
Muscularis externa (circular layer, longitudinal layer, oblique layer)
Serosa

Small Intestine

Identify the following features and sketch and label what you see in the space below (Photos 390c–391b):

Intestinal villi
Mucosa (folded into plicae circulares; lined with simple columnar epithelium with microvilli)
Submucosa (areolar connective tissue)
Muscularis externa (circular layer, longitudinal layer)
Serosa

Large Intestine

Identify the following features and sketch and label what you see in the space below: (Photos 392a and 392b)

Mucosa (simple columnar epithelium, goblet cells)
Submucosa (areolar connective tissue)
Muscularis externa (circular layer, longitudinal layer)
Serosa (fibrous connective tissue)

Liver

The liver's structural and functional units are called **lobules.** Each lobule is a roughly hexagonal structure consisting of plates of **liver cells.** The liver cell plates radiate outward from a **central vein** running within the longitudinal axis of the lobule. Between the liver cell plates are enlarged capillaries called **liver sinusoids.** Identify the following features and sketch and label what you see in the space below (Photos 393a and 393b):

Lobules
Liver cells
Central vein
Sinusoids

Name _____

Lab 25: Anatomy of the Digestive System

◆ **Practice**

1. What is the difference between a GI tract organ (alimentary canal organ) and an accessory organ?

2. For the list of organ functions below choose the best digestive process description.

    A. ingestion
    B. propulsion
    C. mechanical digestion
    D. chemical digestion
    E. absorption
    F. defecation

_____ Saliva is secreted by the parotid gland into the oral cavity.

_____ The stomach wall churns food prior to passing it through the pyloric sphincter as chyme.

_____ Emulsified fatty materials move into the lacteals within villi of the small intestine.

_____ Indigestible residues are eliminated from the GI tract through the anal canal.

_____ Peristaltic contractions of smooth muscle in the small intestine propel chyme distally in the small intestine.

_____ You place food in your mouth when you eat.

3. Choose the organ that best fits the descriptions below:

   A. cecum
   B. gall bladder
   C. jejunum
   D. sublingual gland
   E. uvula
   F. duodenum
   G. ileum
   H. liver
   I. sigmoid colon
   J. rectum
   K. teeth
   L. tongue
   M. stomach
   N. large intestine
   O. esophagus
   P. pancreas

   _____ The outer layer of this GI tract organ is the adventitia

   _____ Teniae coli pull this long tube into pockets called haustra

   _____ Mucosal folds are called rugae

   _____ Manufactures and secretes bile

   _____ Muscular structure moves food around the mouth

   _____ Thin-walled sac stores bile

   _____ Positioned between the duodenum and ileum

   _____ Secretes a fluid that begins the process of carbohydrate digestion in the mouth

   _____ Prevents food substances from moving into the nasopharynx

   _____ Receives bile and pancreatic juice to continue digestive processes

   _____ S-shaped region of the large intestine

   _____ The first region of the large intestine that receives food residues from the ileum

   _____ Involved in mastication

4. What are the functions of the mesenteries? Name and list the functions of the specialized mesenteries attached to the stomach.

DIGESTIVE PROCESSES

Lab 26

Objectives

1. List specific digestive enzymes and the products upon which they function
2. Identify the locations of action of specific digestive enzymes
3. Explain the conditions under which some enzymes must function
4. Demonstrate the function of bile in fat digestion

Materials

- Water bath
- Ice

For each lab group:
- hot plate
- 1,400 ml beaker, hot water bath w/ boiling chips
- 1,400 ml beaker for water ice bath
- wax pencil (2)
- test tube rack with test tubes
- spot plate
- test tube holders (4)
- transfer pipets
- tub of soapy water for dirty test tubes
- 24 well tray with dropper
- pipets of each of the following:
  - Benedict's solution
  - Lugol's solution
  - 1% maltose

- 1% boiled starch
- 1% amylase
- Distilled water
- 1% pancreatin
- Litmus cream (Heavy cream stirred together with litmus powder)
- Vegetable oil
- Bile salts (in a glass vial)

Protein Digestion Demonstration test tubes (covered with paraffin):
- A1—trypsin + whole boiled egg white + water at 37°C
- A2—trypsin + chopped boiled egg white + water at 37°C
- B1—trypsin + whole boiled egg white + ice water
- B2—trypsin + chopped boiled egg white + ice water
- C1—Boiled trypsin + whole boiled egg white + water at room temperature
- C2—Boiled trypsin + chopped boiled egg white + water at room temperature
- D1—Whole boiled egg white + distilled water
- D2—Chopped boiled egg white + distilled water
- E1—Chopped boiled egg white + distilled water
- E2—chopped boiled egg white + hydrochloric acid (HCl)
- E3—chopped boiled egg white + water + pepsin
- E4—chopped boiled egg white + HCl + pepsin

INTRODUCTION

As foodstuffs move through the gastrointestinal tract, they are mechanically and chemically digested into their component building blocks, or monomers. It is these small, simpler nutrients that can then be absorbed into the blood or lymph. The agents that accomplish the chemical digestion are enzymes secreted by various body cells. Because these enzymes can also function outside of the body, they can be used in the laboratory to demonstrate digestive processes. Before proceeding with this lab, review the Laboratory Guidelines & Safety.

CONCEPTS

The four major types of food nutrients are carbohydrates, proteins, lipids and nucleic acids. **Carbohydrates** can be consumed as starch from plants such as potatoes, grains and fruits. Milk contains carbohydrates in the form of lactose. Our body cells are not able to metabolize complex carbohydrates and so must use enzymes to hydrolyze them into simpler monosaccharides. **Proteins,** obtained from meats, dairy products, legumes and grains are composed of amino acids, which can be used for metabolism and also for synthesis of new proteins. **Lipid** building blocks are fatty acids and glycerols, both of which may be metabolized. When not immediately needed for energy, however, fatty acids and glycerols are recombined as triglycerides and stored as fat. **Nucleic acid** monomers are nucleotides, which can later be reassembled into new DNA or RNA. Table 26.1 shows the enzymes that work on each of the food components and where the chemical digestion occurs.

Table 26.1
Enzyme Action on Foods

◆◆◆

Food Material	Enzyme	Site of Chemical Digestion	Absorption of Nutrients
Carbohydrates	Salivary amylase secreted from salivary glands	Oral cavity	Absorption occurs in small intestine through the villi
	Pancreatic amylase from pancreas	Small intestine	Absorbed through villi
	Brush border enzymes from small intestine	Small intestine	Absorbed through villi
Proteins	Pepsin from stomach precursor	Stomach	Absorbed through villi in small intestine
	Trypsin from pancreas precursor	Small intestine	Absorbed through villi
	Chymotrypsin from pancreas precursor	Small intestine	Absorbed into blood capillaries through villi
	Carboxypeptidase from pancreas precursor	Small intestine	Absorbed into blood capillaries through villi
	Brush border enzymes	Small intestine	Absorbed into blood capillaries through villi
Fats	Bile from liver	Small intestine	Emulsifies fats for easier chemical digestion
	Pancreatic lipase	Small intestine	Absorption through villi into lacteals
Nucleic Acids	Pancreatic ribonuclease	Small intestine	Absorbed into blood capillaries through villi
	Pancreatic deoxyribonuclease	Small intestine	Absorbed into blood capillaries through villi
	Brush border enzymes	Small intestine	Absorbed into blood capillaries through villi

◆ Exercise 26.1 *Test the Action of Digestive Enzymes in Various Conditions*

Positive and Negative Controls

1. Sugar
   In this lab we will use Benedict's solution to test for the presence of sugar (maltose or glucose). When adding blue Benedict's solution to a test substance, the following color changes indicate the presence of sugar in increasing amounts:

   blue (–)
   yellow / white (++)

   We will establish the validity of the use of Benedict's solution as a test agent by setting up a positive and negative control.
   a. Using separate droppers, place ½ ml of maltose and ½ ml of Benedict's solution in a test tube marked with a +.
   b. Using separate droppers, place ½ ml of starch solution and ½ ml of Benedict's solution in a test tube marked with a –.
   c. Use a test tube holder to transfer the test tubes into a beaker of boiling water and heat for five minutes.
      ▶ What is the color of the solution/precipitate in tube +? _____

      ▶ What is the color of the solution/precipitate in tube –? _____

      ▶ Explain why this control experiment validates the use of Benedict's solution to detect the presence of sugar.

2. Starch
   In this lab we will use Lugol's solution to test for the presence of starch. When starch reacts with Lugol's solution the precipitate is a dark blue to black color. A negative test results in a yellow to brown color.
   a. On one spot plate depression place a drop of starch and on another depression place a drop of maltose. *Use two different droppers!*
   b. Place a drop of Lugol's solution onto the starch and maltose.
      ▶ What color change occurred in the drop of starch?

      ▶ What color change occurred in the drop of maltose?

      ▶ Explain how this control experiment validates the use of Lugol's solution to detect the presence of starch.

Digestive Enzyme Tests

1. Digestion of Fats

   The pancreas secretes pancreatic juice, which contains enzymes that can digest fats, proteins, nucleic acids and carbohydrates. The enzyme you will use in this test is pancreatic lipase, which digests fats to glycerol and fatty acids. The fatty acids produced change the color of blue litmus to red.

   a. Use a dropper to place 1 ml of litmus cream into two test tubes. Litmus cream is a heavy (fat) cream to which powdered litmus has been added.

   b. Place one tube in a 40°C water bath and the other in an ice water bath. Wait 5 minutes to allow the tubes to reach the appropriate temperature.

   c. Use a dropper to add 1 ml of pancreatic juice to each tube and return each tube to its water bath until you note a color change in one of the tubes. This may take up to 20 minutes. The color change may be subtle.

   ▶ What color change did you notice in the 40°C tube?

   ▶ What color change did you notice in the 0°C tube?

   ▶ Explain how temperature affects fat digestion.

2. Digestion of Starch

   In this experiment you will use a solution of salivary amylase, which begins starch digestion in the mouth.

   a. Using separate droppers, add 1 ml of starch solution and 0.5 ml of amylase solution to a test tube and swirl gently.

   b. Wait for 1 minute and remove 1 drop of the mixture with a glass rod to the depression of a spot plate. Test for the presence of starch with Lugol's solution.

   c. At 20 second intervals test samples of the mixture until you no longer obtain a positive test result for starch. Use the glass rod to stir the mixture after each test.

   d. After the first negative test, test the remaining mixture for the presence of glucose. Add 1 ml of Benedict's solution to the test tube and heat in a boiling water bath for 5 minutes.

   ▶ How long did it take for the starch to be digested?

   ▶ What was the result when you tested the mixture for the presence of sugar after the starch was digested?

   ▶ What is the meaning of your result?

3. Effect of Bile on Fats

While bile does not contain any enzymes, it is extremely important in the digestion of fats due to its emulsifying effects. Emulsification is the breaking down of large globules into smaller ones so that the fats can be uniformly distributed in fluid. Lipases then have access to the greater surface area of small fat globules and can more readily act upon them.

a. Place 5 ml of water into each of two test tubes. Add two "shakes" of bile salts into one of the tubes. Label the test tubes to distinguish them.

b. Place a drop of vegetable oil that has been colored with a fat soluble dye into each tube.

c. Stopper each tube and shake vigorously for 30 seconds.

d. Place both tubes into a test tube rack and leave undisturbed for 10 minutes. Fat or oil that has been broken down into smaller droplets will remain in an emulsion. Fat that has not emulsified will float on the surface of the water.

▶ What difference do you detect in the appearance of the mixture in the two test tubes?

▶ How does emulsification of lipids by bile aid in the digestion of fat?

4. Demonstration: Digestion of Proteins

This experiment has been prepared for you in advance. You will examine the results and make conclusions about the effects of enzyme, temperature and pH on the digestion of protein found in boiled egg whites.

Digestion of protein in the small intestine by the enzyme trypsin:

a. Test tubes marked "A" contain egg white combined with trypsin in water at 37°C (body temperature). Tube A1 contains whole egg white while tube A2 contains chopped egg. What is the appearance of the egg and fluid in each of the test tubes?

▶ Does the enzyme work better on the whole egg or the chopped egg? Why?

b. Test tubes marked "B" contain egg white combined with trypsin in ice water. Tube B1 contains whole egg while tube B2 contains chopped egg.

▶ Is there a difference between the amount of digestion in tubes B1 and B2?

Compare tubes B1 and B2 with tubes A1 and A2. In which tubes does more digestion occur? _____

▶ What does this tell you about the importance of temperature on the action of trypsin?

c. Test tubes marked "C" contain egg white combined with boiled trypsin. C1 contains whole egg while C2 contains chopped egg. Compare tubes C1 and C2 with A1 and A2.

▶ In which tubes does more digestion occur? _____

▶ What was the effect of boiling trypsin prior to adding it to the egg protein?

d. Test tubes marked "D" contain egg white in distilled water with no trypsin. Tube D1 contains whole egg while tube D2 contains chopped egg. Compare tubes D1 and D2 with tubes A1 and A2.

▶ In which tubes does more digestion occur? _____

▶ Is the enzyme necessary for protein digestion? _____

▶ Explain why test tubes D1 and D2 are good controls for this experiment.

e. Digestion of protein in the stomach by the enzyme pepsin:
Test tube E1 contains chopped egg, distilled water and no enzyme.
Test tube E2 contains chopped egg, hydrochloric acid (HCl) and no enzyme.
Test tube E3 contains chopped egg, water and pepsin.
Test tube E4 contains chopped egg, HCl and pepsin.

▶ In which tube has the most digestion occurred? _____

▶ Explain the effect of pH on the action of pepsin.

▶ Which tubes are controls in this experiment? What condition does each tube control for?

Name _____

Lab 26: Digestive Processes

◆ **Practice**

1. What enzymes digest carbohydrates?

2. In what locations are these enzymes secreted?

3. What are the end products of starch and carbohydrate digestion?

4. What enzyme in the stomach digests proteins?

5. What enzymes in the small intestine digest proteins?

6. What are the end products of protein digestion?

7. How does bile assist the digestion of fats?

8. What enzyme found in pancreatic juice digests fats?

9. Where do pancreatic enzymes digest food substances?

10. What are the end products of fat digestion?

11. Which enzyme(s) require a low pH for optimal function?

12. What is the advantage of the optimal pH for enzyme function?

13. What is the advantage of the pancreas secreting enzyme precursors into the duodenum rather than secreting active enzymes?

ANATOMY OF THE URINARY SYSTEM

Lab
27

Objectives

1. Identify the organs of the urinary system on a torso model
2. Identify specific structures in a kidney on a model and in a pig kidney
3. Trace the blood supply of the kidney in a model and on diagrams
4. Explain structural differences in the male and female urethra
5. Describe the structure of nephrons and identify the parts of the nephron on a microscope slide

Materials

- Torso model
- Kidney model
- Pig kidneys
- Dissecting instruments
- Disposable gloves
- Microscope slide of kidney

INTRODUCTION

The urinary system is an elegant and intricate mechanism for processing and renewing blood plasma. Waste substances are eliminated while desirable material is reabsorbed. A minimum amount of fluid which contains waste products is excreted as urine, and additional water can be excreted as a system for fluid balance. Electrolytes are also regulated through the urinary system. pH balance depends, in part, on the elimination of H^+ ions in urine. In this lab you will examine the organs of the urinary system and identify the parts of the nephron, the site of urine formation.

CONCEPTS

I. Gross Anatomy of the Urinary System

The urinary system includes all organs involved with the production, storage and elimination of urine. Urine formation is the result of interaction between the kidney and its blood supply. The **kidneys** are positioned retroperitoneally on the posterior abdominal wall, deep to the floating ribs. Perched on the superior surface of each kidney is the **adrenal gland,** which participates in urine formation by regulating blood pressure. Blood supply and the ureter enter the kidneys at a medial indentation, the **renal hilus.** The ureter is a muscular transport tube that conveys urine to the **urinary bladder** located in the pelvic cavity. Urine exits the urinary bladder through the **urethra.**

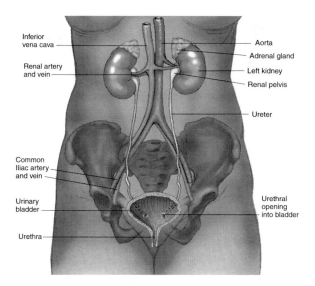

FIGURE 27.1 ◆ Urinary System Organs

◆ **Exercise 27.1** *Identify Urinary Organs in a Torso Model*

Practice reviewing functions of the urinary organs as you identify them in a torso model.

A. Gross Anatomy of the Kidney (Figure 27.2 and Photo 394)

Each kidney is covered by a protective fibrous **renal capsule.** The kidney itself has an outer cortical region and a deeper medullary region. The **renal cortex** is paler in color. The **renal medulla** is actually a series of dark pyramid shaped structures called the **renal pyramids.** A band of cortex exists between each pyramid and is called the **renal column.** A kidney **lobe** consists of a renal pyramid, the adjacent renal columns and the cap of renal cortex. Urine is formed by structures within the renal pyramids and is released to a hollow system of spaces within the kidney. The tips of the pyramids, called **papillae,** project into small spaces called **minor calyces.** The minor calyces converge in two or three **major calyces,** which come together in a funnel shaped **renal pelvis.** From the renal pelvis urine moves into the ureter.

Blood supply to the kidney is an integral part of urine formation. Blood arrives at the kidney through the **renal artery** off of the **abdominal aorta.** The renal artery divides into five **segmental arteries,** which each separate into several **lobar arteries** located in the renal columns. From the lobar arteries branch interlobar arteries. These give rise to **arcuate arteries** that arch over the superficially located bases of the renal pyramids. Smaller **interlobular arteries** continue into the cortex and feed into arterioles.

Blood returns from the kidney through veins with the same names as the arteries. Capillary blood drains into **interlobular veins** in the cortex and then into **arcuate veins** that parallel arcuate arteries. From there blood moves into **interlobar veins,** however there are no lobar or segmental veins. The interlobar veins drain into the **renal vein** which empties into the **inferior vena cava.**

◆ **Exercise 27.2** *Identify Structures in a Kidney Model*

Identify the following structures in a model of the human kidney (Photo 394): cortex; medullary pyramids; renal columns; minor calyces; major calyces; renal pelvis; ureter; arteries: renal, segmental, lobar, interlobar, arcuate, interlobular; veins: interlobular, arcuate, interlobar, renal

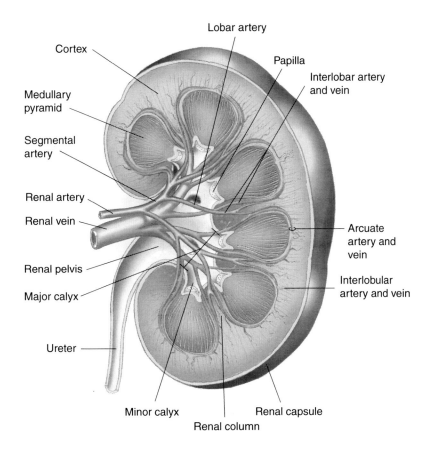

Lobar artery

Cortex

Papilla

Interlobar artery
and vein

Medullary
pyramid

Segmental
artery

Renal artery

Renal vein

Arcuate
artery and
vein

Renal pelvis

Major calyx

Interlobular
artery and vein

Ureter

Minor calyx

Renal capsule

Renal column

FIGURE 27.2 ◆ Structures in the Kidney

◆Exercise 27.3 *Identify Structures in a Pig Kidney*

Use Photo 395 to assist with this dissection.

1. Obtain a preserved pig kidney and dissecting instruments. Observe the external transparent **renal capsule.**
2. Find the **ureter, renal vein,** and **renal artery** at the **hilus** region.
3. Make a cut through the longitudinal axis (frontal section) of the kidney and identify the following structures and spaces:
   cortex
   medullary pyramids (renal pyramids)
   renal columns
   renal pelvis
   renal papilla
   major calyx
   minor calyx

4. If the preserved kidney is injected with latex, follow the renal blood supply from the renal artery to the cortex and back to renal vein.

B. Ureters

Urine travels through the **ureters** retroperitoneally to the urinary bladder. The lumen of the ureter is lined with a stretchy transitional epithelium surrounded by smooth muscle. Peristaltic contractions of the muscle assist urine movement. The outer adventitia is a fibrous connective tissue.

C. Urinary Bladder

Ureters enter the **urinary bladder** at the posterior and inferior aspect. This angle of entry causes the openings of the ureters to constrict as urine fills the bladder. The bladder rises superiorly to accommodate increasing volumes of urine. The muscular wall of the bladder, called the **detrusor muscle,** exhibits **rugae** much like the stomach. Transitional epithelium comprises the mucosa, and fibrous adventitia covers the external surface. The urethra exits the bladder at the inferior aspect. The openings for the two ureters and the urethra form a triangular region called the **trigone.** Bacteria can enter the bladder through the urethra and establish residence in the trigone region, causing a persistent bladder infection. The junction between the urinary bladder and urethra is controlled by the **internal urethral sphincter,** which is under autonomic control.

D. Urethra

The mucosa of the **urethra** begins as transitional epithelium near the bladder, but becomes stratified squamous epithelium closer to the **external urethral orifice** (opening). In females, the urethra is only about 1.5 inches long and ends anterior to the vaginal orifice. This position and short length increase the vulnerability to infection in sexually active women. The male urethra is about 8 inches long and has three regions. The **prostatic urethra** courses through the prostate gland after exiting the bladder. Prostatic hypertrophy commonly occurs as men age and can pinch and occlude the prostatic urethra. The **membranous urethra** is the short length within the urogenital diaphragm, or body cavity wall. The **spongy** (penile) **urethra** is positioned within the penis. The male urethra serves double duty as a urinary and reproductive structure since it passes urine and semen.

◆ **Exercise 27.4** *Identify Urinary Structures in a Cat Specimen*

1. Use Photos 412 and 413 to assist you in identifying urinary structures. Open the abdominopelvic cavity on the cat. Gently lift out the intestines and move them to one side to expose the posterior abdominal wall. The **kidneys** are located retroperitoneally, and may be encapsulated in fat. Gently remove the fat to expose the surface of the kidney.

2. Examine the **renal artery and vein** coming off the abdominal aorta and inferior vena cava. Notice how the vessels enter the kidney at the **renal hilus** along with the **ureters.**

3. Use the scissors to carefully cut open the superficial **renal capsule** of the kidney. Gently separate it from the surface of the kidney. Use the scalpel to make a cut through the frontal plane of the kidney. You should be able to identify the **renal pelvis** and the **renal cortex** and **medulla.**

4. Trace the ureter from the renal hilus to its termination at the **urinary bladder.** Do not cut the ureter. Note the position of the junction between the ureters and the urinary bladder. Use scissors to cut open the urinary bladder, and note the folds of **rugae** on the inner wall. These allow the expansion of the bladder as it fills with urine. Note the muscular wall of the urinary bladder. This muscle is called the **detrusor** muscle.

5. Find the **urethra** as it exits the inferior aspect of the urinary bladder. Dissect out the urethra to the external urethral opening. If your cat is a male, look for the **prostate gland** through which the urethra travels.

◆◆◆

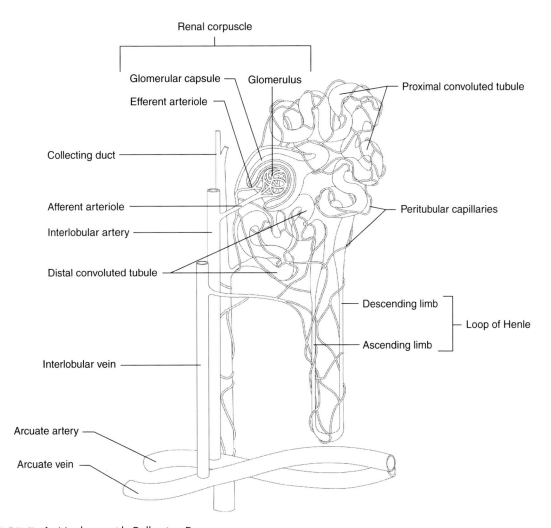

FIGURE 27.3 ◆ Nephron with Collecting Duct

II. Microscopic Anatomy of Kidney: Nephrons (Figure 27.3, Photos 396a and 396b)

The kidney contains over one million nephrons, the functional part of the kidney. The **nephron** consists of the renal tubule and its blood supply. The **renal tubule** is a long tube that begins at a closed end and empties urine into a **collecting duct.** The tube has four regions: **glomerular capsule, proximal convoluted tubule (PCT), loop of Henle, distal convoluted tubule (DCT).** The glomerular capsule surrounds a capillary bed called the **glomerulus** and receives fluid and solutes as a **filtrate** from the blood plasma. The glomerulus together with the glomerular capsule is called the **renal corpuscle.** The filtrate moves through the rest of the tubule and is processed into urine along the way. Some substances from within the filtrate are reabsorbed out of the tubule and into the interstitial space.

The blood supply surrounding the renal tubule is like a package service. It delivers materials to one region and then picks up modified materials from another region. The interlobular arteries within the cortex branch into **afferent arterioles** that supply the glomerular capillary beds. The glomerulus is unique in that it is also drained by an arteriole, the **efferent arteriole.** The diameter of the afferent arteriole is greater than that of the efferent arteriole, which creates increased blood pressure within the glomerulus. The increased pressure forces fluid and solutes out of the glomerular capillary beds into the glomerular capsule to become the filtrate. The efferent arterioles feed into the **peritubular capillaries** that surround the renal tubule. Fluid and solutes that have been reabsorbed by the tubule into the interstitial space diffuse into the peritubular capillaries and rejoin blood circulation. In this way liters of fluid are processed and returned to circulation, and only a small amount of fluid becomes urine. The peritubular capillaries empty into the interlobular veins.

◆ **Exercise 27.5** *Examine a Microscope Slide of Nephrons*

1. Begin your observation of the microscope slide under low power. Identify a **glomerulus** which appears as a ball of tightly packed material containing many small nuclei, and the **glomerular capsule.** (Photos 396a, 396b)

2. Notice that the **renal tubules** are cut at various angles. Also try to differentiate between the thin-walled **loop of Henle** portion of the tubules and the cuboidal epithelium of the proximal convoluted tubule, which has dense microvilli. Sketch and label what you see in the space below.

Name _____

Lab 27: Anatomy of the Urinary System

◆ Practice

1. Place these blood vessels in order from aorta to inferior vena cava:

   a. interlobular vein
   b. interlobular artery
   c. lobar artery
   d. afferent arteriole

   e. interlobar vein
   f. interlobar artery
   g. peritubular capillaries
   h. efferent arteriole

   i. segmental artery
   j. renal vein
   k. glomerulus
   l. arcuate vein

   m. arcuate artery
   n. renal artery

   aorta → _____ → _____ → _____ → _____ → _____ → _____ → _____ →
   _____ → _____ → _____ → _____ → _____ → _____ → _____

   → inferior vena cava

2. How long is the average male urethra? How long is the average female urethra?

3. Explain how the length and position of the female urethra increase vulnerability to infection.

4. Label the parts of the kidney in the diagram:

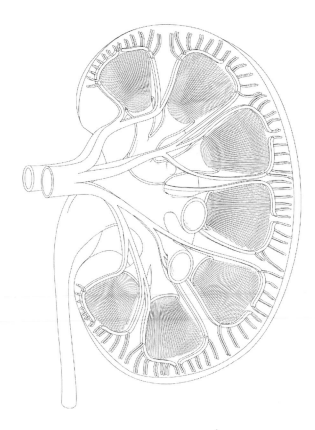

5. Label the parts of the nephron in the diagram:

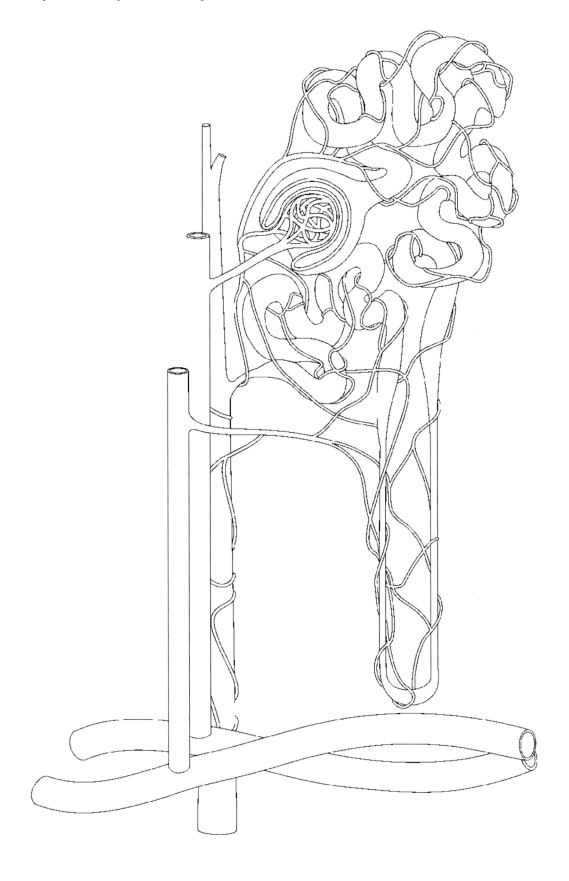

REPRODUCTIVE ANATOMY AND FEMALE REPRODUCTIVE CYCLES

Lab 28

Objectives

1. Identify structures of the male and female reproductive systems on cat specimens
2. Identify membranes and structures associated with a pig ovary
3. Trace the pathway followed by a sperm from its site of formation to its destination
4. Trace the pathway followed by an egg from its site of formation to its destination
5. Identify microscopic structures of the male and female reproductive systems
6. Describe the events of the ovarian cycle and menstrual cycle
7. Explain the role of hormones in the ovarian and menstrual cycles

Materials

- Cats specimens; males and females
- Pig ovaries
- Dissecting trays
- Dissecting instruments
- Disposable gloves
- Microscope slides
  - seminiferous tubules
  - ovary (with 1°, 2°, and vesicular follicles)

INTRODUCTION

Sexual reproduction requires genetic material from two parents to come together in the process of fertilization and form a unique new set of genetic instructions for the offspring. The sex cells, or **gametes,** that unite their nuclei in fertilization must each have half of the normal number of chromosomes (**haploid**), so that the new individual will have the correct number (**diploid**). The male and female reproductive systems produce gametes for the purpose of sexual reproduction. Male gametes are **sperm** and female gametes are commonly called **eggs,** but more accurately called **ova** (singular is **ovum**). The male reproductive system is specialized to deliver sperm to the female reproductive anatomy. Within the female the sperm may unite with and fertilize the egg. Because the sperm delivers only genetic material to the egg, the egg is responsible for providing the cellular machinery that will respond to genetic instructions. After fertilization the diploid cell performs a rapid series of cell divisions to create additional cells that become an embryo and later a fetus. The female reproductive system is uniquely designed to house and nourish the fetus as it grows, and to support the nutritional needs of the baby after it is born.

CONCEPTS

I. Male Reproductive Anatomy

A. External Genitalia

The function of the **penis** is to deliver sperm to the female vagina. To enter the vagina the penis must become erect. Three columns of erectile tissue in the shaft of the penis provide the ability to become a penetrating organ. Erectile tissue is composed of connective tissue and smooth muscle laden with vascular spaces served by arterioles. Upon arousal, arteriole sphincters relax and permit blood to fill vascular spaces, causing the penis to become rigid. The columns of erectile tissue include a single, midventral **corpus spongiosum** and two dorsal **corpora cavernosa.** The distal end of the penis is composed of an enlarged region of corpus spongiosum, the **glans.** The

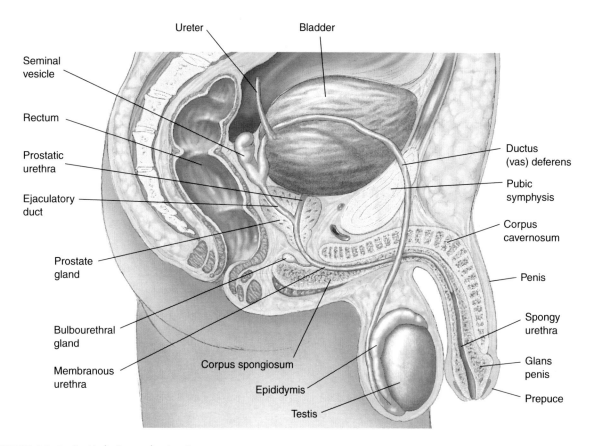

FIGURE 28.1 ◆ Male Reproductive Structures

spongy urethra courses through the center of the corpus spongiosum. Located in the center of each of the corpora cavernosa is an artery that serves the arterioles of the erectile tissue. Skin loosely covers the penis shaft. The unattached skin at the glans is called the **prepuce,** or foreskin. The penis is anchored to the body wall by the root, which is comprised of two **crura** (singular is **crus**) and the **bulb** of the penis.

Sperm are produced by structures within the paired **testes.** Because sperm development occurs at a narrow temperature range below that of body temperature, the testes are suspended in an external cutaneous sac, the **scrotum.** A muscle within the scrotal wall, the **dartos muscle,** will wrinkle if temperature drops too low, so that air will become trapped in the spaces next to the scrotum to be warmed. Another muscle, the **cremaster muscle,** surrounds each testis and draws the testis closer to the body wall when temperature lowers. Blood and nerve supply to the testes are wrapped in a **spermatic cord** that extends from the external inguinal canal which passes through the body wall. The **testicular artery** carries hot arterial blood toward the testis. The **pampiniform plexus** is a network of much cooler testicular veins that surround the testicular artery and draw off much of the heat before the blood in the artery arrives at the testis.

B. **Duct System**

Sperm are produced within and travel through ducts to exit the reproductive tract. Long thin coils of **seminiferous tubules** comprise the testes and are the site of sperm production. **Interstitial cells** are positioned around the seminiferous tubules and secrete testosterone to support the development of sperm. Sperm exit the seminiferous tubules via the short **tubulus rectus** into the network of ducts at the posterior aspect of the testis called the **rete testis.** From here sperm enter the **epididymis,** which is another coiled system of tubes that form a cord-like structure from the superior to inferior aspects of the posterior testis. Sperm are stored within the epididymis until ejaculated. Upon ejaculation, smooth muscle contraction within the walls of the epididymis propel the sperm distally into the **ductus deferens,** also called the vas deferens. The ductus deferens lies adjacent to the epididymis on

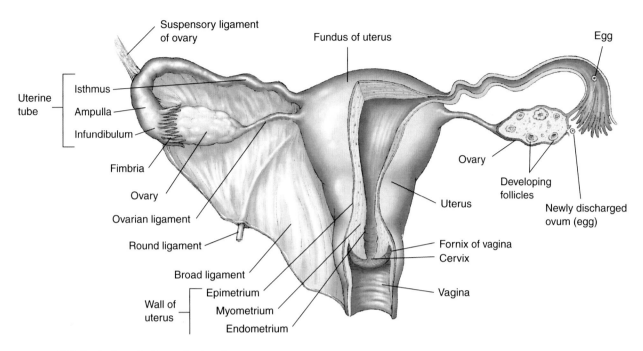

FIGURE 28.2 ◆ Female Reproductive Structures

the posterior testis, then ascends through the spermatic cord into the pelvic cavity. It continues along the supero-lateral urinary bladder and ends in a widened **ampulla** at the posterior surface of the bladder. Muscle contractions propel the sperm through the ductus deferens and into a short **ejaculatory duct** which joins the **prostatic urethra.** Here sperm enter a sticky fluid, **semen,** that is ejaculated through the **membranous** and **spongy urethra** and out the **external urethral orifice.**

C. Glands

Two glands contribute to the composition of semen. The **seminal vesicle** lies anterior to the ampulla of the ductus deferens and ejects a viscous fluid that comprises about 60% of the semen. This alkaline fluid contains a coagulating enzyme to create a thick medium for sperm transport, and fructose sugars for sperm nourishment. The **prostate gland** secretes additional fluid with nutrients and activating enzymes for the sperm.

A third gland, the **bulbourethral gland,** located in the urogenital diaphragm or body wall, does not contribute to semen. Instead, it secretes a thick, alkaline fluid into the spongy urethra to counteract the acidic residues in the urethra from urine.

II. Female Reproductive Anatomy

A. External Genitalia

Female external genitalia are collectively called the **vulva.** The **mons pubis** is a fatty mounded region anterior to the pubic symphysis. The **labia majora** are two folds of skin that define the lateral borders of the vulva. Medially located are two **labia minora** that are homologous to the ventral penis. The labia minora surround a region called the **vestibule,** which contains the external urethral orifice and the vaginal orifice posteriorly. The **clitoris,** covered by a **prepuce,** is positioned at the anterior apex of the vestibule. The clitoris is homologous to the penis and is composed of erectile tissue. The exposed portion of the clitoris is the **glans.**

B. Duct System

The **vagina** is the site of sperm deposition for sexual reproduction. It is a short passageway from the **external vaginal orifice** to the cervix of the uterus. It is also the passageway for menstrual flow and for the birth of a baby. The internal vaginal walls are ridged to stimulate the male penis. The mucosa is covered with an acid mantle that inhibits growth of pathogens. Projecting into the superior vagina is the rounded surface of the cervix of the uterus. The spaces within the vagina around the cervix are the fornices.

The **uterus** is a thick walled sac that supports the growth of a baby. From the narrow **cervix,** the uterus widens into a **body** and has a rounded **fundus** at the superior surface. The wall of the uterus consists of three layers: the endometrium, myometrium and perimetrium. The **endometrium** is the simple columnar mucosal lining of the uterus lumen. A rich vascular supply creates a cushioned and nourishing environment for an embryo before the embryo grows a functioning placenta. If no embryo is present, the inner lining of the endometrium is sloughed off as the monthly menstrual flow. Smooth muscle layers compose the **myometrium,** and are responsible for labor contractions. The external **perimetrium** is the visceral peritoneum.

Extending from the superolateral aspects of the uterus are two **uterine tubes,** also called fallopian tubes. The narrow region of the uterine tube at its connection with the uterus is the **isthmus.** A widened **ampulla** exists at the distal end and connects to the open ended **infundibulum.** Extending from the edges of the infundibulum are finger-like **fimbriae.** Movement of the fimbriae creates a current that draws the female egg into the uterine tube. The sperm meet the egg in the uterine tube, and one sperm may fertilize the egg in this duct. The uterine tube mucosa contains cilia that create a current that moves the egg toward the uterus.

C. Ovaries

Ovaries are the site of egg production in females. The ovaries are attached to the lateral uterine wall via a stout ovarian ligament. The infundibulum of the uterine tube opens in close proximity to the ovary. Within the cortex of the ovary are located spherical follicles that contain **oocytes,** which will become ova. Under the influence of follicle stimulating hormone, the smallest **primordial follicles** become **primary follicles,** which become slightly larger **secondary follicles.** Eventually the secondary follicles become **vesicular follicles** and bulge out the side of the ovary. Each month a spike of luteinizing hormone stimulates ovulation of a vesicular follicle and the oocyte is released. The oocyte gets swept into the infundibulum of the uterine tube by the fimbriae currents.

D. Membranes and Round Ligament

The uterus and ovary are supported within the pelvic cavity by the **broad ligament.** The **mesometrium** is the mesentery that creates the bulk of the broad ligament inferior to the ovarian ligament and superior to the vagina. The fundus of the uterus is secured to the labia majora by the round ligaments, which course from the fundus through the inguinal canal.

◆ Exercise 28.1 *Identify Reproductive Structures in a Cat Specimen*

Male (Photo 413)

1. Identify the **scrotum** of the cat. Make an incision through the wall of the scrotum to expose the **testes** within. The muscular wall of the scrotum is called the **dartos muscle.**
2. Remove the testis from the scrotum. The tough membrane covering the surface of the testis is the **tunica vaginalis.** Examine the **epididymus** on the posterior surface.
3. The epididymus is continuous with the **ductus deferens,** which travels superiorly through the spermatic cord. Dissect out the **spermatic cord** through the abdominal wall and trace the ductus deferens to its termination at the **prostate gland.** You may have to lift the intestines out of the way.
4. Dissect out the **penis,** and trace the urethra from the urinary bladder to the penis. At the root of the penis find the **bulbourethral glands** on either side.

Female (Photo 414)

1. Lift the intestines out of the lower abdominal and pelvic region. Gently move the urinary bladder to one side. Deep to the urinary bladder, and anterior to the descending colon, you will find the **vagina** where it joins the inferior pelvic wall at the base of the tail.
2. The vagina is continuous superiorly with the **body of the uterus.** Unlike in humans, the body of the uterus splits into two **uterine horns,** one on each side of the abdominal cavity. These structures are analogous to the uterine tubes in humans. The uterine horns may still be attached to the abdominal wall by the membranous **broad ligament.**

3. At the superior end of each uterine horn you will find a small, bumpy **ovary** connected to the uterine horns by a thin **ovarian ligament.** If your dissection technique is very good, you may be able to trace, from the end of the uterine horn, the tiny cat **uterine tube** with the slightly enlarged **infundibulum** touching the ovary.

III. Microscopic Anatomy of Reproductive System

◆ Exercise 28.2 *Identify Structures in Microscope Slides of Seminiferous Tubules and Ovary*

Male (Photos 397a–397d)

Examine a slide of *seminiferous tubules* and draw and label the following structures:

Lumen of seminiferous tubule

Spermatozoa (sperm within the seminiferous tubules will be immature)

Spermatogenic cells

Interstitial cells (Leydig cells)

Female

Examine a slide of *ovary* and identify the following structures. **Primordial follicles** will be surrounded by a single layer of squamous **follicle cells. Primary follicles** have at least two layers of either cuboidal or columnar **granulosa cells.** Follicles which have a space, called the **antrum,** surrounding the oocyte are **secondary follicles** until they bulge from the surface of the ovary, when they are called **vesicular follicles.**

Sketch and label what you see in the space below. (Photos 398a and 398b, Figure 28.3)

Primordial follicles

Primary follicles

Secondary follicles

Vesicular (Graafian) follicles

    Oocytes

    Granulosa cells

    Antrum

    Corpus luteum

IV. Female Reproductive Cycles

A. Ovarian Cycle

The 28-day **ovarian cycle** is regulated by **follicle stimulating hormone (FSH)** and **luteinizing hormone (LH)**. Events of the ovarian cycle are listed below:

Follicular phase (days 1–14)—period of follicular and oocyte growth

1. Primordial follicle becomes a primary follicle
   - Oocyte stimulates the primordial follicle to become the primary follicle
2. Primary follicle becomes the secondary follicle
   - The single layer of **follicular cells** of the primary follicle divide and become multiple layers of **granulosa cells**
   - Granulosa cells stimulate oocyte to grow and form the zona pellucida membrane
   - Formation of a fluid-filled **antrum** around the oocyte indicates a secondary follicle

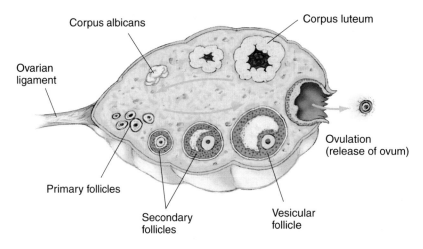

FIGURE 28.3 ◆ Internal Structure of Ovary

3. Secondary follicle becomes a vesicular follicle
   - The antrum expands and follicle bulges outside of ovary
   - The oocyte is attached to granulosa cells only on one side

Ovulation—after a stimulating spike of LH, a single (usually) vesicular follicle ruptures the wall of the ovary and ejects the oocyte

Luteal Phase (days 15–28)—after ejection of the oocyte the granulosa cells convert into estrogen and progesterone secreting cells and form the **corpus luteum,** which lasts for 14 days.

1. After 14 days, if the oocyte is not fertilized the corpus luteum degenerates into scar tissue called the **corpus albicans**
2. If the oocyte is penetrated by a sperm it becomes an **ovum** until the two nuclei join in **fertilization.** The resultant **zygote** can then become an **embryo.** The embryo produces **Human Chorionic Gonadotropin (HCG)**, which stimulates the corpus luteum to continue to secrete hormones for 12 weeks. After 12 weeks the **placenta** takes over hormone production and the corpus luteum becomes a corpus albicans. (HCG is the substance detected in positive pregnancy tests.)

B. **Menstrual (Uterine) Cycle**

The 28-day **menstrual cycle** is regulated by estrogens and progesterone. It occurs simultaneously with the ovarian cycle. Events of the menstrual cycle phases are listed below:

Menstrual phase (days 1–5)—the uterus sheds its superficial endometrial lining

Proliferative phase (days 6–14)

1. Deep layers of the endometrium regenerate the superficial layer
2. The mucus plug in the cervix thins to improve access for sperm
3. Estrogen levels increase; a sudden drop in estrogen causes ovulation on day 14

Secretory phase (days 15–28)

1. Progesterone levels rise from corpus luteum, causing the endometrium to form a secretory mucosa to support an embryo
2. Estrogen levels rise slightly
3. The cervical plug thickens
4. If no embryo secretes HCG the corpus luteum degenerates, progesterone and estrogen decline, and the menstrual phase begins

Lab 28: Anatomy of Reproductive System

◆ Practice

1. Trace the path of sperm from their site of production to the external urethral orifice. Include every structure through which the sperm pass.

2. What two glands contribute to the production of semen?

3. What general substances are secreted by the glands in question 2?

4. Which hormones regulate the ovarian cycle?

5. Describe each of the following follicles:

   a. primordial

b. primary

c. secondary

d. vesicular

6. Which hormone stimulates ovulation?

7. Trace the path of the oocyte from its site of production to its exit during the menstrual phase from the vaginal orifice. List every structure through which the egg passes.

8. What is the function of the corpus luteum?

9. What happens to the corpus luteum after 14 days if:

    a. the oocyte is fertilized and becomes an embryo

    b. the oocyte is not fertilized

10. What chemical signal is produced by the embryo that affects the corpus luteum?

11. What would be the result to a pregnancy if the placenta forms improperly and cannot take over hormone production for the corpus luteum? At how many weeks of development would you expect this "changing of the guard"?

12. List the events of the menstrual cycle and include the relative amount and roles of estrogen and progesterone.

GENES AFFECT ANATOMY AND PHYSIOLOGY

Lab 29

Objectives

1. Define terms related to heredity
2. Use Punnett squares to determine probability of genotype and phenotype
3. Calculate probabilities for sex-linked genotypes and phenotypes
4. Demonstrate the difficulty in eliminating recessive alleles within a population
5. Demonstrate how some traits may be determined by multiple genes
6. Demonstrate the effects of genes and the environment in determining risk of diabetes

Materials

- Two opaque containers per lab group
- Red and white kidney beans with same shape and size
- Lifestyle slips of paper

INTRODUCTION

Your anatomy and physiology is a direct result of the genes within your body cells. Each new individual is the product of the union between a sperm and an egg. These gametes have half of the genetic material of a normal body cell, so that when they join together they create a single cell with the correct number of chromosomes. The process of passing our genes to our offspring is a fascinating topic of study. This lab introduces the basic concepts that underlie heredity.

CONCEPTS

I. Terminology

Table 29.1
Terminology in Genetics

◆◆◆

Diploid	The condition of body cells that have two of each chromosome number. One set of chromosomes came from the sperm and the other set came from the egg.
Haploid	The condition of gametes that have only one of each chromosome number. When two haploid nuclei join in fertilization they form a diploid cell.
Autosome	Chromosomes # 1–22; each body cell will have two of each autosome (**homologous chromosomes**)
Sex Chromosome	Each body cell has two sex chromosomes: XX designates female and XY designates male; one sex chromosome came from the sperm and one came from the egg.
Allele	A gene on a chromosome that matches a gene on a homologous chromosome; each body cell has two copies of every gene, although the genes may code for different traits (e.g. there are two copies of the gene for earlobe shape, but one may code for an attached earlobe and the other may code for an unattached earlobe)
Homozygous	The alleles for a particular gene are the same
Heterozygous	The alleles for a particular gene are different
Expression	The result of transcribing and translating a specific gene; the appearance of certain traits is the result of the synthesis of specific proteins (e.g. melanin)
Dominant	An allele that is always expressed when present (homozygous or heterozygous)
Recessive	An allele that is only expressed when homozygous; its expression is masked by the presence of a dominant allele in a heterozygous condition
Genotype	The alleles that are present on the chromosomes
Phenotype	The expression of the alleles; recessive alleles are not expressed in the heterozygous condition
Incomplete Dominance	Alleles that are equally expressed when present in a heterozygous condition

---

II. Punnett Squares

◆ **Exercise 29.1** *Use Punnett Squares to Determine Genotype and Phenotype Probability in Offspring*

A. Punnett Squares and Complete Dominance

If you know the genotypes of the parents, you can use a Punnett square to determine the probability of the offspring having any particular genotype or phenotype. The simplest Punnett square shows each parent donating one allele. For example, we can draw a square showing the cross between a curly haired parent and a straight haired parent:

1. In this example curly hair is a dominant trait, designated by "C."
   The gene for straight hair is recessive and is designated by "c."
2. Thus, the straight-haired parent must have a genotype of cc.
   ▶ Why?

The curly-haired parent could have a genotype of CC or Cc. In this example the curly-haired parent has the genotype Cc.

Therefore, the cross is shown like this: Cc (curly) x cc (straight)

3. The genotype from one parent is drawn across the top of the square, and the genotype from the other parent is drawn down the left side, one allele per box.

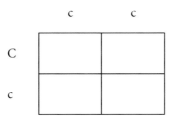

4. Each letter across the top and down the side of the square corresponds to the gene for hair curliness/straightness that is donated to the offspring in a gamete. One gamete from one parent is combined with one gamete from the other parent to provide two copies of the gene, or allele, for hair curliness in the offspring. Each box in the square represents a 25% probability (one box of four is 25%) for the offspring. To determine the possible genotypes, fill in each column with the letter above, and each row with the letter to the left.

	c	c
C	Cc	Cc
c	cc	cc

5. To determine the probability of each *genotype,* add up the number of boxes for each genotype and multiply times 25%.

   Cc is in 2 boxes. 2 × 25% = 50%

   cc is in 2 boxes. 2 × 25% = 50%

6. To determine the probability of each *phenotype,* do the same calculation as in #5 for each phenotype possibility.

   Cc will be expressed as curly hair. Cc is in 2 boxes. 2 × 25% = 50%

   cc will be expressed as straight hair. cc is in 2 boxes. 2 × 25% = 50%

7. Thus, each child will have a 50% chance of having the genotype Cc, and have 50% chance of having the genotype cc. If the child is Cc, her phenotype is curly hair. If the child is cc, her phenotype is straight hair.

B. Punnett Squares and Incomplete Dominance

In incomplete dominance, each allele is equally expressed. For this example we will use plants that have alleles for red flowers and for white flowers. We will use "R" for red and "r" for white.

1. If a homozygous red flower plant is crossed with a homozygous white flower plant, the Punnett square would look like this:

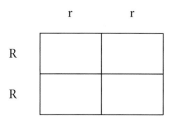

2. Fill in the square with the possible gamete combinations for the plant offspring.
3. What percentage of the offspring will have the genotype Rr ? _____
   (Hint: Your answer should be 4 × 25%.)
4. In incomplete dominance:
   RR will be expressed as red flowers
   rr will be expressed as white flowers
   Rr will be expressed as *pink* flowers

The R and r are equally expressed, causing an equal amount of red and white color in the flowers that appear pink.

C. Punnett Squares and Sex-linked Traits

Sex-linked traits are those traits that are coded for by genes on either the X or Y chromosome. These traits deserve special attention because women do not have a Y chromosome and will never express a Y-linked trait, and men only have one Y chromosome and so a Y-linked trait will always be expressed. Also, though women have two X chromosomes, men have only one and will always express an X-linked trait.

   ▶ When will women express a recessive X-linked trait?

1. In this example, a woman who is a carrier, heterozygous, for colorblindness (XX^c) marries a man who is not colorblind (XY). The Punnett square to show possible offspring genotypes looks like this:

<table>
<tr><td></td><td>X</td><td>Y</td></tr>
<tr><td>X</td><td></td><td></td></tr>
<tr><td>X^c</td><td></td><td></td></tr>
</table>

2. Fill in the Punnett square above.
   ▶ What percentage of girls will be carriers of colorblindness?

   ▶ What percentage of boys will be colorblind?

III. Selection Against Recessive Traits

◆ **Exercise 29.2** *Attempt to Eliminate Recessive Alleles from a Population*

If recessive traits are often negative, and indeed even fatal, why do they continue to persist in the population? In this exercise you will use beans to model a fictitious system in which individuals expressing a recessive trait are not allowed to reproduce.

A rabbit breeder prefers silky smooth coats on his rabbits to rough coats. He gives away all rabbits with rough coats to local children, while breeding the silky rabbits to show.

1. Obtain an opaque container filled with an equal number of white and brown beans. The brown beans signify genes for a silky coat and the white beans represent the rough coat gene.
2. Have one lab partner hold the container and other lab partner draw out all of the beans in pairs without looking, and place the pairs on the table. Because the rough coat is a recessive trait, it is only expressed when the genotype is homozygous for the rough coat, or as white-white in the beans. Brown-white pairs represent heterozygous carriers that are phenotypically silky, and brown-brown rabbits are homozygous silky. Record in Table 29.2 the total number of individuals in this first generation (pairs of beans), the number of recessive alleles (single white beans), number of homozygous recessive (pairs of like color) and heterozygous (pairs of different color) individuals.
3. Remove all of the white-white pairs to a separate pile on the table, then replace all of the remaining beans back in the container. Hold the container closed and shake it to represent recombination of gametes during reproduction of all individuals. (We all know how quickly rabbits can reproduce.)
4. Repeat step 2, drawing out all of the beans in pairs and recording numbers in Table 29.2. Eliminate all homozygous, white-white, individuals to a separate pile and replace the remaining beans in the container to be recombined. Continue this process for five generations.

Table 29.2
Results for Exercise 29.2

◆◆◆

Generation	# Individuals (# of pairs of beans)	# Recessive alleles (total # of white beans)	# Homozygous recessive individuals (# of pairs of like-color beans)	# Heterozygous individuals (# of pairs of different-color beans)
1				
2				
3				
4				
5				

▶ Were you able to completely eliminate the rough coat rabbits from each new generation?

▶ Why?

◆◆◆

IV. Complex Human Traits

The inheritance of most human traits is controlled by more than one gene. For example, multiple genes contribute to height, weight, eye color and skin color. All of these characteristics have a wide variety of possible phenotypes, and environmental influences increase the possibilities. In the next two exercises you will explore models of inheritance using multiple genes and the effect of environmental factors.

◆ **Exercise 29.3** *Demonstrate the Effect of Multiple Alleles for Hair Color*

1. Traits Determined by Multiple Genes
   In this activity the brown beans represent a dark hair allele and the white beans represent a blond hair allele. The beans should be equally distributed between two opaque containers, one labeled "Mother" and one labeled "Father." Each student should participate in this activity.

2. A single gene trait can be modeled by removing one allele from each of the Mother and Father containers. Look around your table at the different allele combinations. You should see three possible combinations:

   brown-brown
   brown-white
   white-white

   If each brown bean represents additional brown color, the brown-brown genotype would have brown hair and the white-white genotype will have blond hair.

3. Now explore the implications of having two genes for hair color. Return all beans to the containers. Draw two beans from each container to represent two hair color alleles from each parent.
   ▶ List the possible genotype combinations with 4 alleles for hair color:

   ▶ How many genotypes are possible? _____

4. Return all beans to the two containers. Draw three beans from each container to represent three hair color alleles from each parent.
   ▶ List the possible genotype combinations with 6 alleles for hair color:

   ▶ How many genotypes are possible? _____

   Each genotype corresponds to a different hair color phenotype.
   ▶ Which multi-gene model is most accurate to depict the numbers of brown hair color phenotypes?

   Realize that while we have illustrated this point using only a brown and blond color, additional genes may contribute red, black and yellow pigments to hair color.

◆ Exercise 29.4 *Demonstrate the Combined Effects of Genes and the Environment on Development of Diabetes*

While hair color can have aesthetic implications, it can be changed with a special shampoo and doesn't affect health of the individual. Some genes, however, may cause a predisposition for specific diseases. In this activity you will model the effects of both genes and environmental factors on the risk of developing type II diabetes.

Note: While they are consistent with current understanding of diabetes risk factors, the scores and number of genes in this activity are chosen for illustration only.

1. Your table should have two containers of beans labeled Mother and Father. From the lab cart draw a slip of paper from each of four bowls labeled "A—Body Weight," "B—Diet," "C—Age," and "D—Exercise." Each slip of paper lists one lifestyle choice and an associated score. Record your "choices" below.

Bowl	Your Characteristic	Your Score
A		
B		
C		
D		

Your Total Lifestyle Score = _____ points

Higher scores generally indicate a higher risk of developing diabetes. However, the risk of developing diabetes also depends on heredity. These genes have not yet been identified by medical professionals, so no one knows what their genetic risk of diabetes is before they choose a lifestyle.

2. Draw three beans from each container, representing three alleles from each parent. Record the color of your beans and then return them to the original container.

Colors: _____   _____

_____   _____

_____   _____

Add 4 points for each brown bean you drew to determine your genetic risk _____

3. Determine your total risk of developing diabetes by adding your lifestyle score to your genetic risk.

Total risk for type II diabetes _____ points

In this illustration, a score of 28 or higher indicates that you have developed type II diabetes. Each student who "suffers" diabetes should write their name on the white board along with their lifestyle score and genetic risk score.

4. Questions
   a. What lifestyle and environmental factors do you know that increase risk of diabetes?

   b. Did any students develop diabetes who had relatively healthy (low) lifestyle scores? What is the meaning of their result?

   c. What is one indication that an individual might have a genetic risk of developing diabetes?

◆◆◆

Name _____

Lab 29: Genes Affect Anatomy and Physiology

◆ **Practice**

1. In guinea pigs, if curly hair (s) is recessive to straight hair (S), what are the percentages of each possible genotype and phenotype for a cross between a curly haired and a homozygous straight haired guinea pig?

2. Male pattern baldness is an X-linked recessive gene. Draw a Punnett square for a cross between a woman who carries the gene and a man with male pattern baldness. What percentage of boys will have pattern baldness? What will the genotypes of the girls be?

3. If hairy feet (H) is dominant to not hairy feet (h), what percentages of each genotype and phenotype will be possible in a cross between two heterozygous individuals?

4. Blood type is an example of incomplete dominance. Each allele is equally expressed. What percentage of each blood type in the offspring would you expect from a cross between an AB mother and an AO father?

5. Define the following terms:

  a. Allele

  b. Homozygous

  c. Autosome

  d. Dominant

e. Genotype

f. Phenotype

INDEX OF TABLES

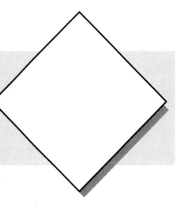

*Table for student data

INDEX OF PHOTOS

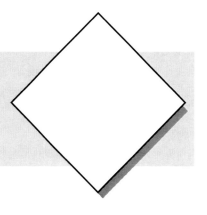

PHOTO ATLAS

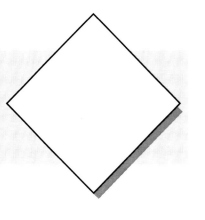

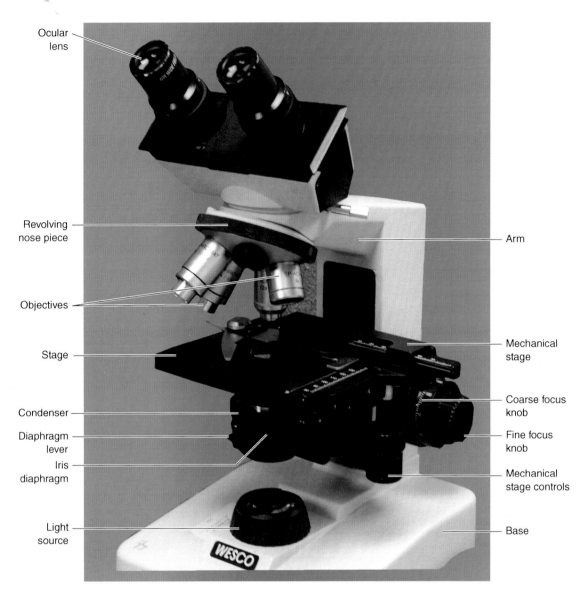

Ocular lens

Revolving nose piece

Objectives

Stage

Condenser

Diaphragm lever

Iris diaphragm

Light source

Arm

Mechanical stage

Coarse focus knob

Fine focus knob

Mechanical stage controls

Base

WESCO

PHOTO 351a ◆ Binocular microscope

Magnification

CHROM
0x / 0.65

Numerical aperture

Lens

PHOTO 351b ◆ Objective lens

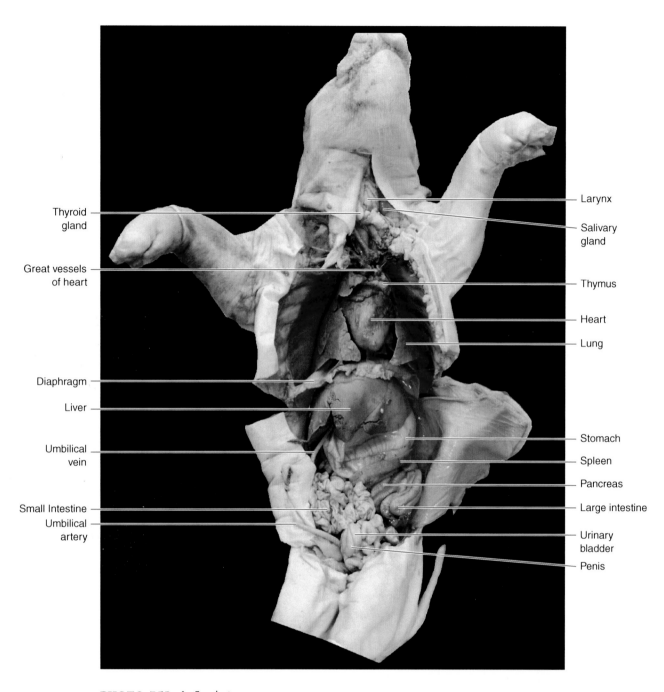

Thyroid
gland

Great vessels
of heart

Diaphragm

Liver

Umbilical
vein

Small Intestine

Umbilical
artery

Larynx

Salivary
gland

Thymus

Heart

Lung

Stomach

Spleen

Pancreas

Large intestine

Urinary
bladder

Penis

PHOTO 352 ◆ Fetal pig

Interphase

Nucleus

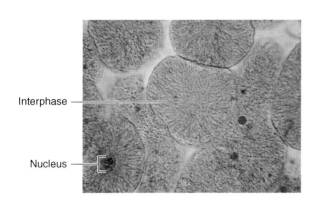

PHOTO 353a ◆ Cell in interphase
(1000x)

Chromosomes
Prophase

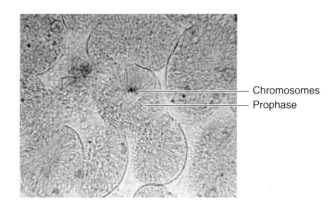

PHOTO 353b ◆ Prophase subphase of mitosis (1000x)

Metaphase

Mitotic
spindle

Chromosomes
aligned on
metaphase plate

Centrioles

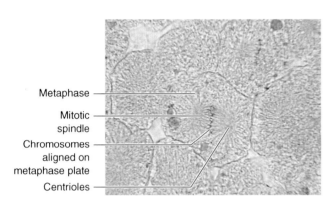

PHOTO 353c ◆ Metaphase subphase
of mitosis (1000x)

Anaphase

Daughter
chromosomes

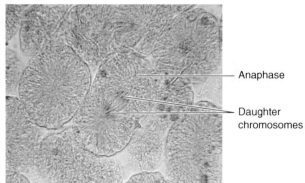

PHOTO 353d ◆ Anaphase subphase of mitosis
(1000x)

Daughter
nuclei

Cleavage
furrow

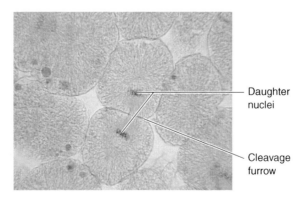

PHOTO 353e ◆ Telophase subphase of mitosis and
cytokinesis (1000x)

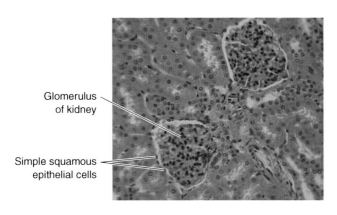

Glomerulus of kidney

Simple squamous epithelial cells

PHOTO 354a ◆ Simple squamous epithelium of renal cortex (400x)

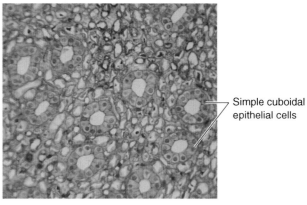

Simple cuboidal epithelial cells

PHOTO 354b ◆ Simple cuboidal epithelium of salivary gland (400x)

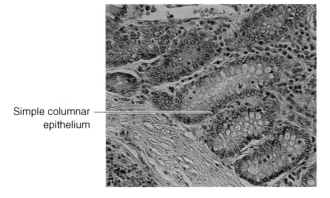

Simple columnar epithelium

PHOTO 354c ◆ Simple columnar epithelium of small intestine (400x)

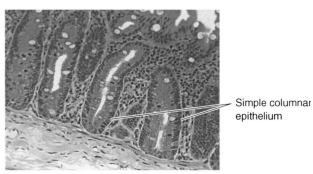

Simple columnar epithelium

PHOTO 354d ◆ Simple columnar epithelium of small intestine (400x)

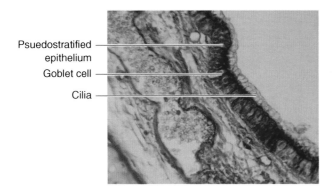

Psuedostratified epithelium

Goblet cell

Cilia

PHOTO 354e ◆ Ciliated pseudostratified epithelium of trachea (400x)

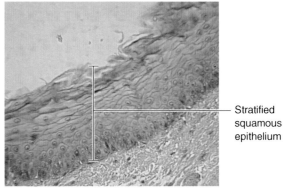

Stratified squamous epithelium

PHOTO 354f ◆ Stratified squamous epithelium of epidermis (400x)

Fibroblast

Elastic
fibers

Collagen
fibers

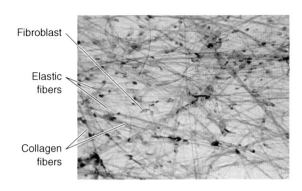

PHOTO 355a ◆ Areolar (loose) connective tissue (400x)

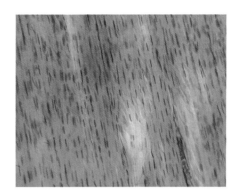

PHOTO 355b ◆ Dense regular connective tissue of tendon (400x)

Collagen
fibers

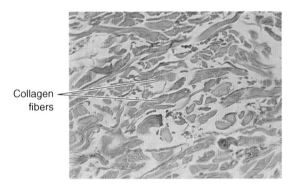

PHOTO 355c ◆ Dense irregular connective tissue of dermis (400x)

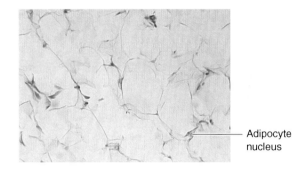

Adipocyte
nucleus

PHOTO 355d ◆ Adipose connective tissue (400x)

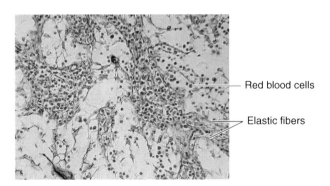

Red blood cells

Elastic fibers

PHOTO 355e ◆ Reticular connective tissue (400x)

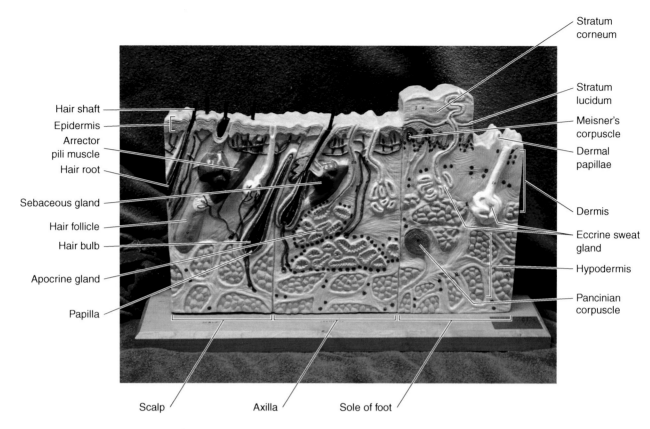

Stratum corneum

Stratum lucidum

Meisner's corpuscle

Dermal papillae

Dermis

Eccrine sweat gland

Hypodermis

Pancinian corpuscle

Hair shaft

Epidermis

Arrector pili muscle

Hair root

Sebaceous gland

Hair follicle

Hair bulb

Apocrine gland

Papilla

Scalp

Axilla

Sole of foot

PHOTO 356a ◆ Skin model

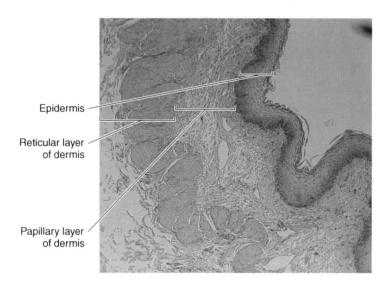

Epidermis

Reticular layer of dermis

Papillary layer of dermis

PHOTO 356b ◆ Thin skin (100x)

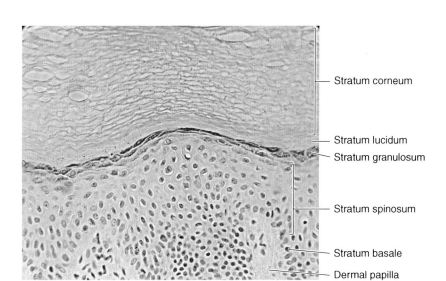

Stratum corneum

Stratum lucidum
Stratum granulosum

Stratum spinosum

Stratum basale
Dermal papilla

PHOTO 357a ◆ Epidermis of thick skin (palm) (400x)

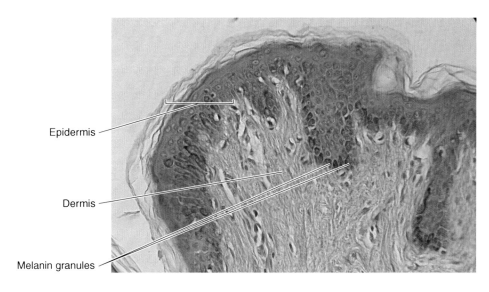

Epidermis

Dermis

Melanin granules

PHOTO 357b ◆ Darkly pigmented skin (100x)

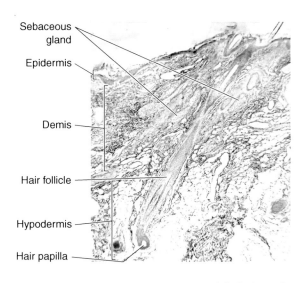

Sebaceous gland
Epidermis
Demis
Hair follicle
Hypodermis
Hair papilla

PHOTO 358a ◆ Hair follicle (400x)

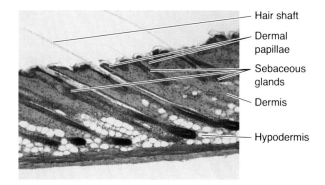

Hair shaft
Dermal papillae
Sebaceous glands
Dermis
Hypodermis

PHOTO 358b ◆ Hair follicles (100x)

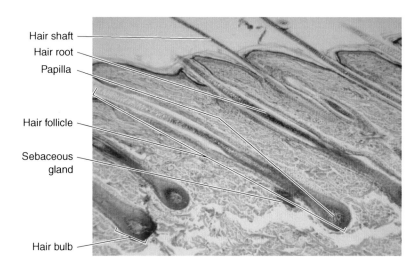

Hair shaft
Hair root
Papilla
Hair follicle
Sebaceous gland
Hair bulb

PHOTO 358c ◆ Hair follicles (100x)

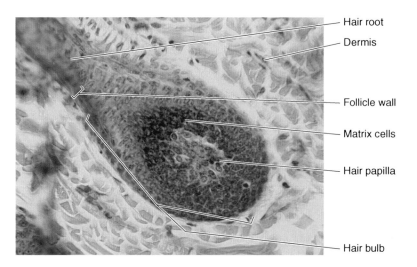

Hair root
Dermis
Follicle wall
Matrix cells
Hair papilla
Hair bulb

PHOTO 358d ◆ Hair bulb (400x)

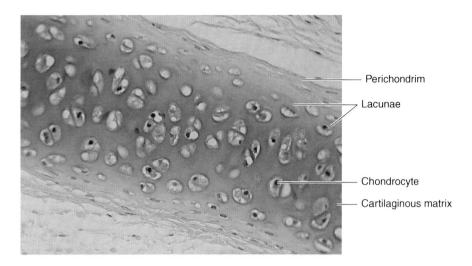

PHOTO 359a ◆ Hyaline cartilage (400x)

Labels: Perichondrim, Lacunae, Chondrocyte, Cartilaginous matrix

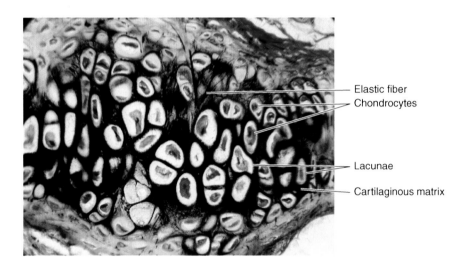

PHOTO 359b ◆ Elastic cartilage (400x)

Labels: Elastic fiber, Chondrocytes, Lacunae, Cartilaginous matrix

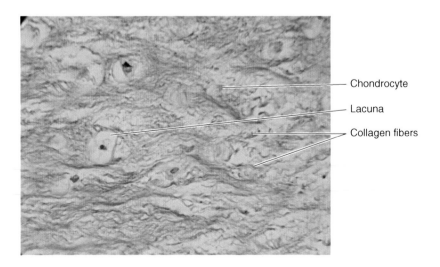

PHOTO 359c ◆ Fibrocartilage (400x)

Labels: Chondrocyte, Lacuna, Collagen fibers

Perforating canal

Osteon

PHOTO 360a ◆ Compact bone (100x)

Lamellae

Central canal

Osteocytes in lacunae

Canaliculi

PHOTO 360b ◆ Osteon of compact bone (400x)

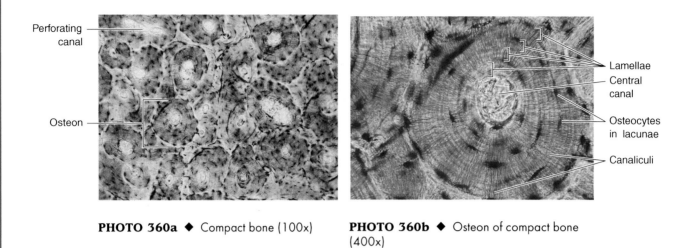

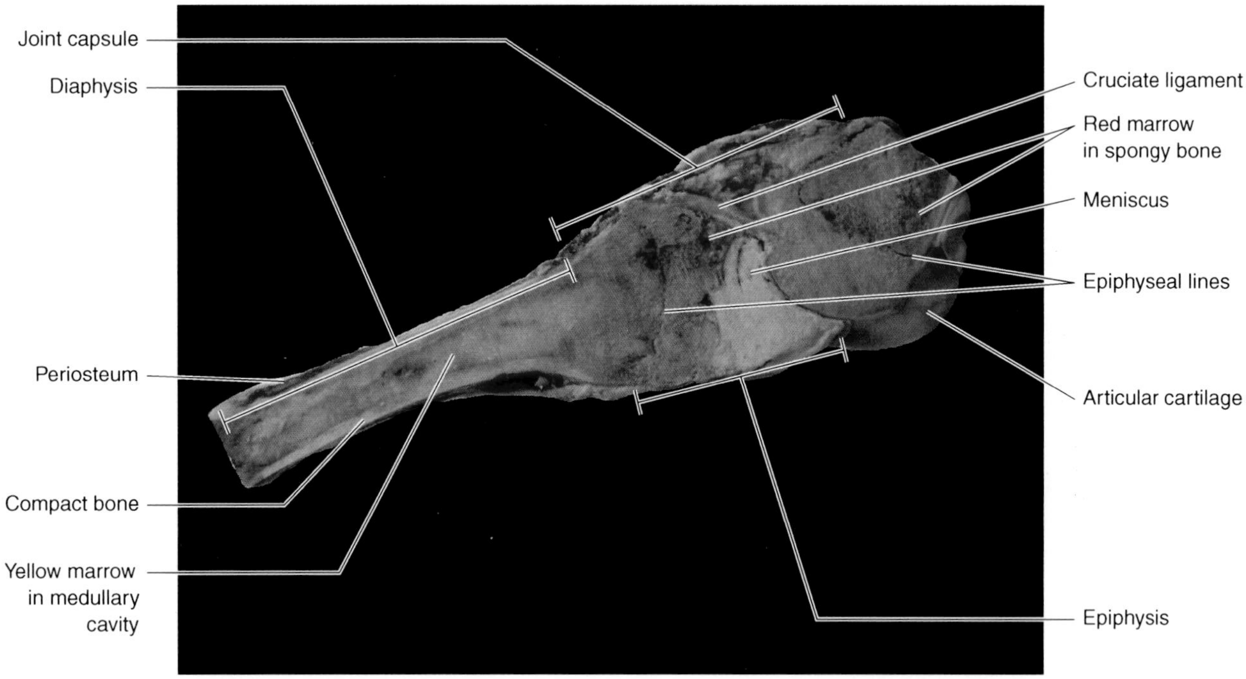

Joint capsule

Diaphysis

Periosteum

Compact bone

Yellow marrow in medullary cavity

Cruciate ligament

Red marrow in spongy bone

Meniscus

Epiphyseal lines

Articular cartilage

Epiphysis

PHOTO 360c ◆ Beef joint

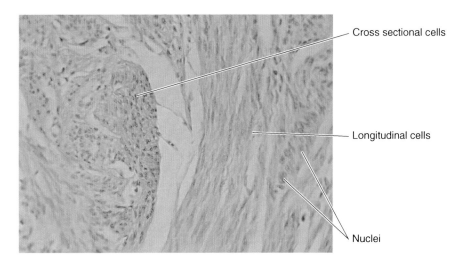

Cross sectional cells

Longitudinal cells

Nuclei

PHOTO 361a ◆ Smooth muscle (400x)

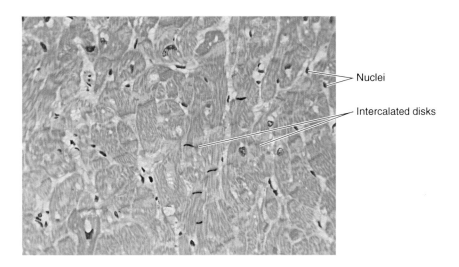

Nuclei

Intercalated disks

PHOTO 361b ◆ Cardiac muscle (400x)

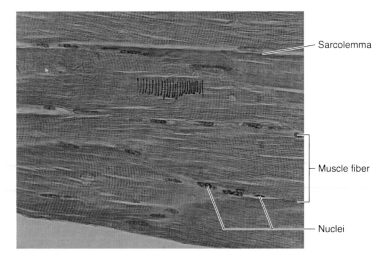

Sarcolemma

Muscle fiber

Nuclei

PHOTO 361c ◆ Skeletal muscle (400x)

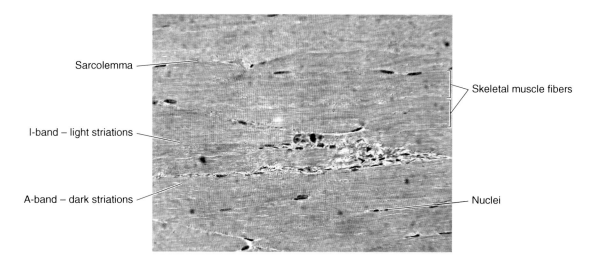

Sarcolemma

I-band – light striations

A-band – dark striations

Skeletal muscle fibers

Nuclei

PHOTO 362a ◆ Skeletal muscle longitudinal section (400x)

Skeletal muscle fibers
lined with endomysium

Fascicle lined with
perimysium

PHOTO 362b ◆ Skeletal muscle cross section (100x)

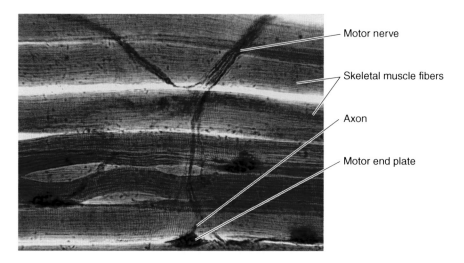

Motor nerve

Skeletal muscle fibers

Axon

Motor end plate

PHOTO 362c ◆ Neuromuscular junction (400x)

Temporalis

Orbicularis oculi

Masseter

Platysma

Sternocleidomastoid

Frontalis

Orbicularis oris

PHOTO 363a ◆ Musculature of head and neck

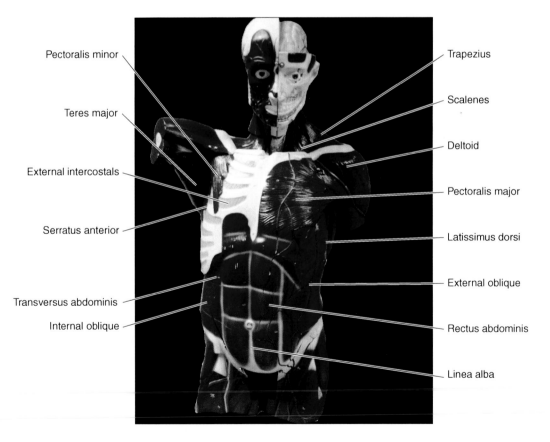

Pectoralis minor

Teres major

External intercostals

Serratus anterior

Transversus abdominis

Internal oblique

Trapezius

Scalenes

Deltoid

Pectoralis major

Latissimus dorsi

External oblique

Rectus abdominis

Linea alba

PHOTO 363b ◆ Musculature of torso

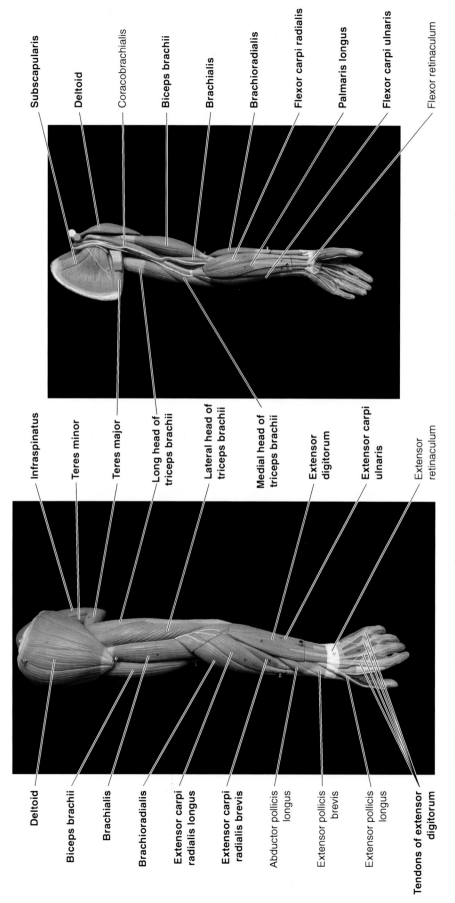

Subscapularis
Deltoid
Coracobrachialis
Biceps brachii
Brachialis
Brachioradialis
Flexor carpi radialis
Palmaris longus
Flexor carpi ulnaris
Flexor retinaculum

PHOTO 364b ◆ Anterior muscles of upper limb

Infraspinatus
Teres minor
Teres major
Long head of triceps brachii
Lateral head of triceps brachii
Medial head of triceps brachii
Extensor digitorum
Extensor carpi ulnaris
Extensor retinaculum

PHOTO 364a ◆ Posterior muscles of upper limb

Deltoid
Biceps brachii
Brachialis
Brachioradialis
Extensor carpi radialis longus
Extensor carpi radialis brevis
Abductor pollicis longus
Extensor pollicis brevis
Extensor pollicis longus
Tendons of extensor digitorum

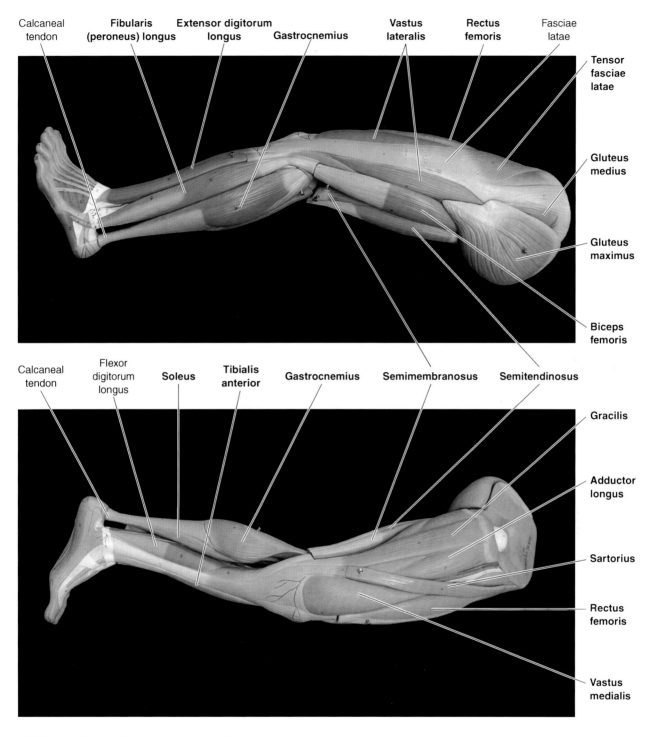

PHOTO 365a ◆ Posterolateral view of lower limb

PHOTO 365b ◆ Medial view of lower limb

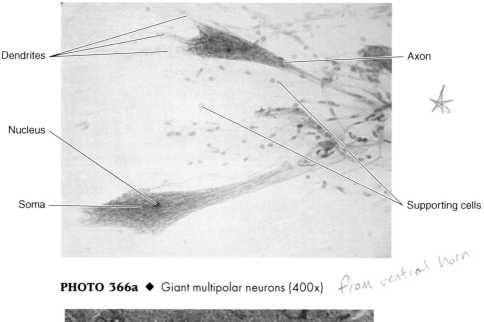

Dendrites

Nucleus

Soma

Axon

Supporting cells

PHOTO 366a ◆ Giant multipolar neurons (400x)   *from ventral horn*

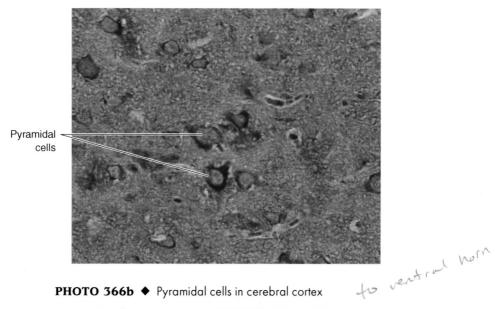

Pyramidal cells

PHOTO 366b ◆ Pyramidal cells in cerebral cortex   *to ventral horn*

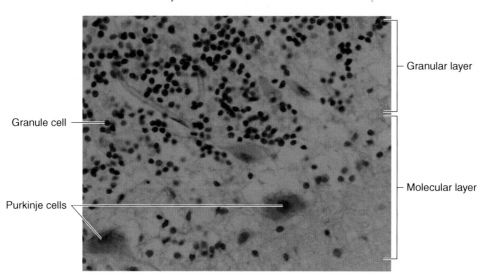

Granule cell

Purkinje cells

Granular layer

Molecular layer

PHOTO 366c ◆ Cerebellum

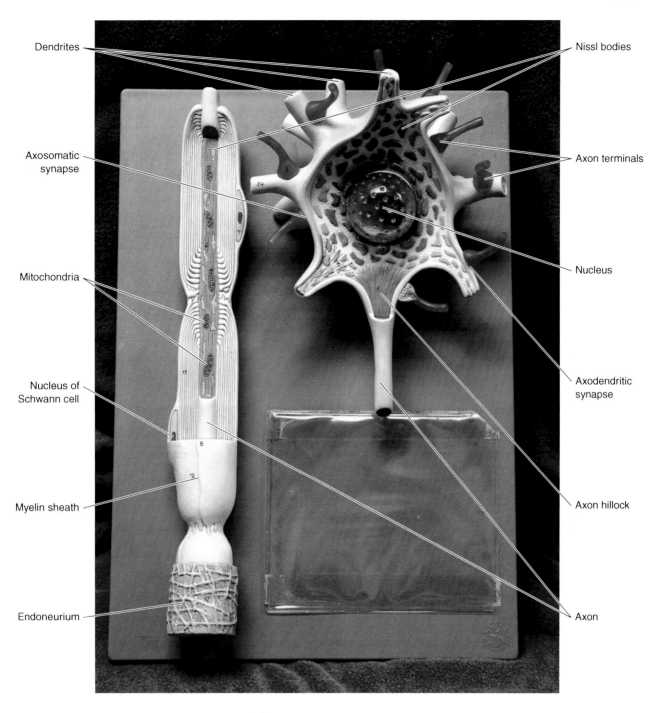

Dendrites

Nissl bodies

Axosomatic synapse

Axon terminals

Mitochondria

Nucleus

Nucleus of Schwann cell

Axodendritic synapse

Myelin sheath

Axon hillock

Endoneurium

Axon

PHOTO 367 ◆ Neuron model

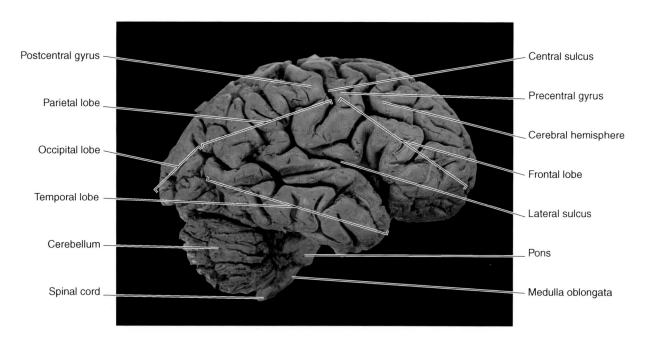

Postcentral gyrus

Parietal lobe

Occipital lobe

Temporal lobe

Cerebellum

Spinal cord

Central sulcus

Precentral gyrus

Cerebral hemisphere

Frontal lobe

Lateral sulcus

Pons

Medulla oblongata

PHOTO 368a ◆ Lateral brain

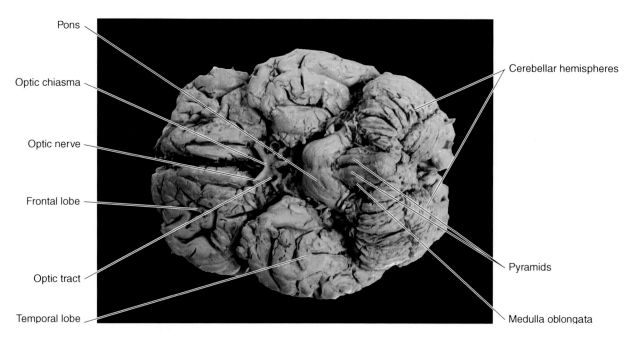

Pons

Optic chiasma

Optic nerve

Frontal lobe

Optic tract

Temporal lobe

Cerebellar hemispheres

Pyramids

Medulla oblongata

PHOTO 368b ◆ Ventral brain

Olfactory tract

Midbrain

Oculomotor
nerve (III)

Trochlear
nerve (IV)

Trigeminal
nerve (V)

Facial
nerve (VII)

Vestibulocochlear
nerve (VIII)

Vagus nerve (X)

Hypoglossal
nerve (XII)

Optic nerve (II)

Optic chiasma

Optic tract

Internal carotid
artery

Cerebral
peduncle

Pons

Abducens
nerve (VI)

Cerebellar
peduncle

Pyramid

Glossopharyngeal
nerve (IX)

Accessory
nerve (XI)

Decussation
of pyramids

Cerebellum

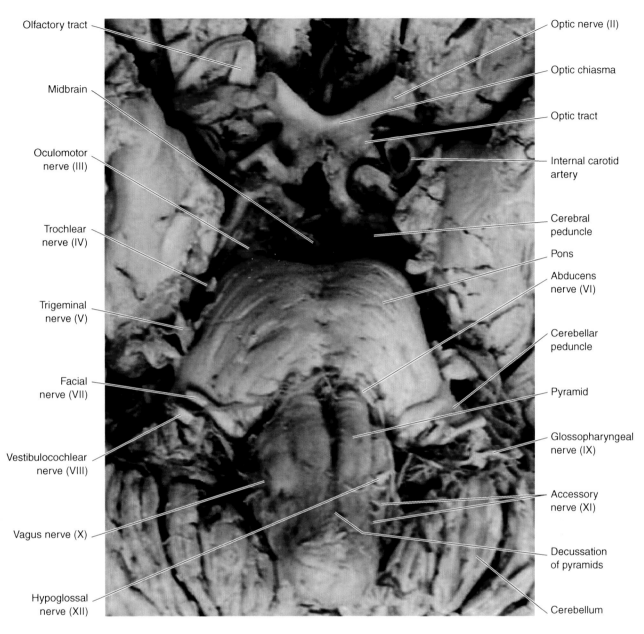

PHOTO 369 ◆ Ventral brainstem

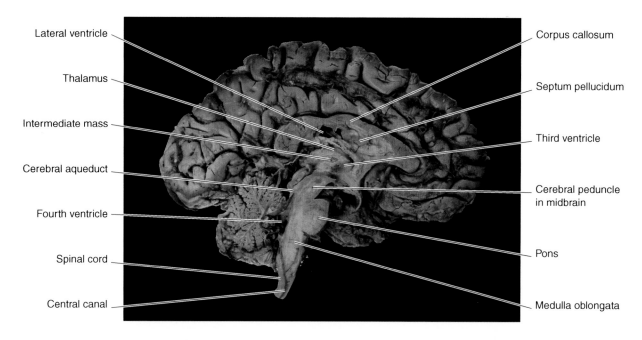

Lateral ventricle

Thalamus

Intermediate mass

Cerebral aqueduct

Fourth ventricle

Spinal cord

Central canal

Corpus callosum

Septum pellucidum

Third ventricle

Cerebral peduncle in midbrain

Pons

Medulla oblongata

PHOTO 370a ◆ Midsagittal brain

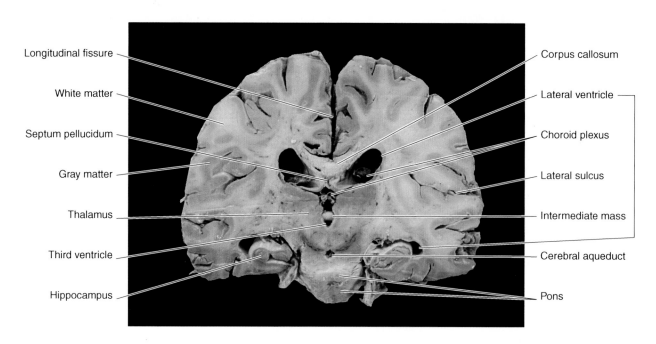

Longitudinal fissure

White matter

Septum pellucidum

Gray matter

Thalamus

Third ventricle

Hippocampus

Corpus callosum

Lateral ventricle

Choroid plexus

Lateral sulcus

Intermediate mass

Cerebral aqueduct

Pons

PHOTO 370b ◆ Frontal (coronal) section through brain

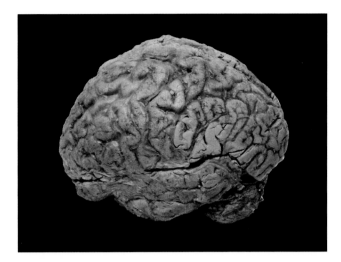

PHOTO 371a ◆ Pia mater on brain

PHOTO 371b ◆ Dura mater on brain

Falx cerebri

Tentorium cerebelli

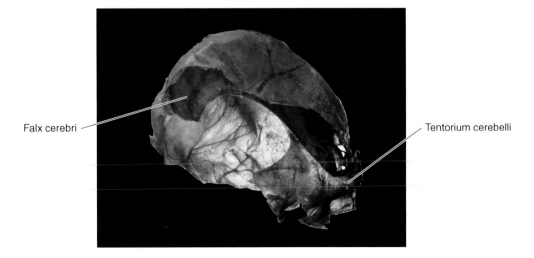

PHOTO 371c ◆ Dura mater

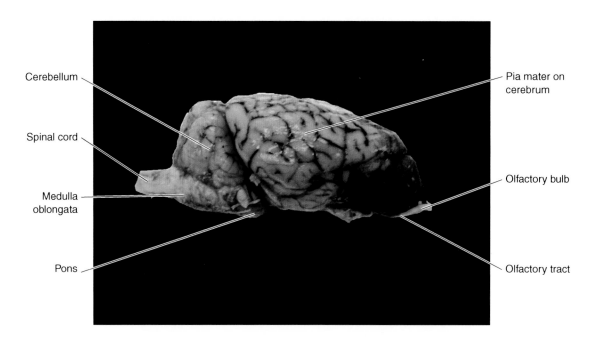

Cerebellum

Spinal cord

Medulla
oblongata

Pons

Pia mater on
cerebrum

Olfactory bulb

Olfactory tract

PHOTO 372a ◆ Lateral sheep brain

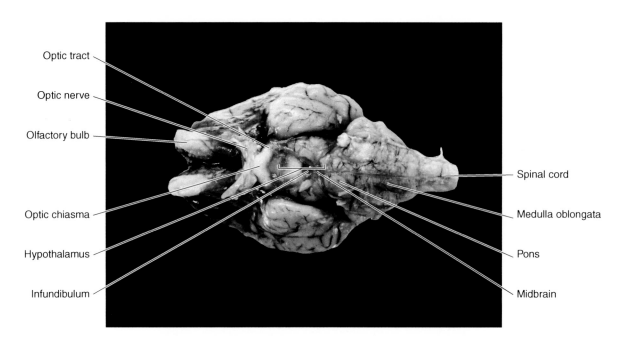

Optic tract

Optic nerve

Olfactory bulb

Optic chiasma

Hypothalamus

Infundibulum

Spinal cord

Medulla oblongata

Pons

Midbrain

PHOTO 372b ◆ Ventral sheep brain

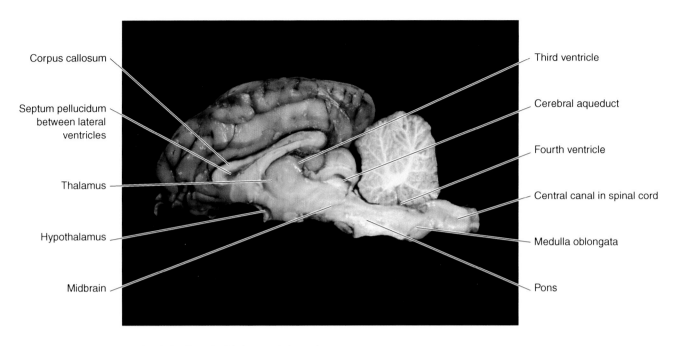

Corpus callosum

Septum pellucidum
between lateral
ventricles

Thalamus

Hypothalamus

Midbrain

Third ventricle

Cerebral aqueduct

Fourth ventricle

Central canal in spinal cord

Medulla oblongata

Pons

PHOTO 373a ◆ Midsagittal sheep brain

Pineal gland

Superior
colliculi

Corpora
quadrigemina

Inferior
colliculi

Cerebral hemisphere

Third ventricle

Fourth ventricle

Cerebellum (reflected)

PHOTO 373b ◆ Dorsal midbrain in sheep brain

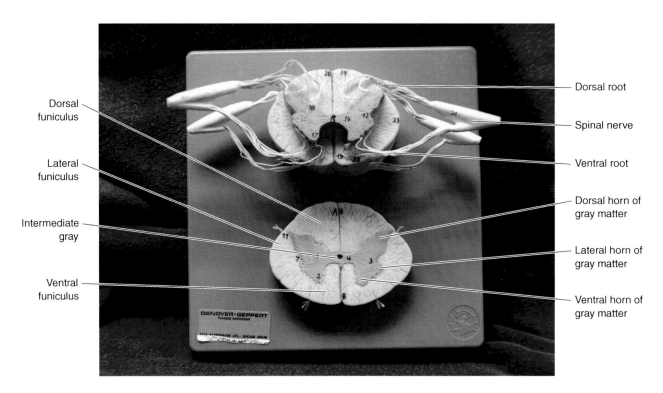

Dorsal funiculus

Lateral funiculus

Intermediate gray

Ventral funiculus

Dorsal root

Spinal nerve

Ventral root

Dorsal horn of gray matter

Lateral horn of gray matter

Ventral horn of gray matter

PHOTO 374a ◆ Spinal cord cross section model

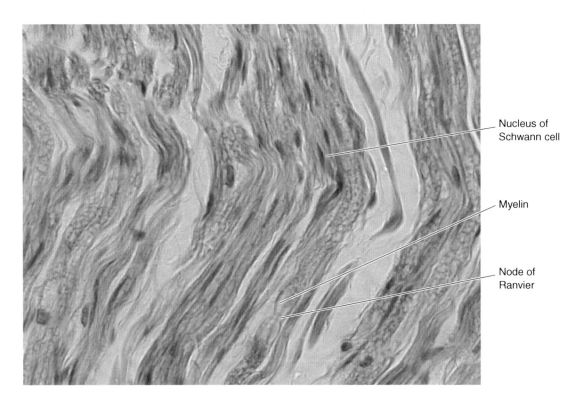

Nucleus of Schwann cell

Myelin

Node of Ranvier

PHOTO 374b ◆ Peripheral nerve longitudinal section (400x)

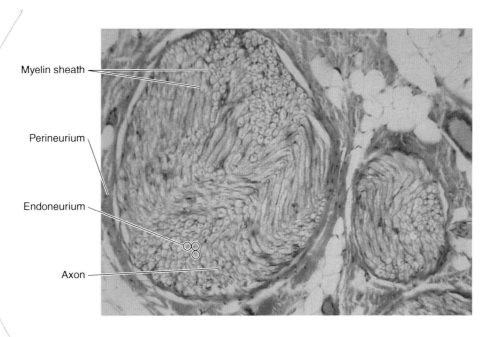

Myelin sheath

Perineurium

Endoneurium

Axon

PHOTO 375a ◆ Peripheral nerve cross section (400x)

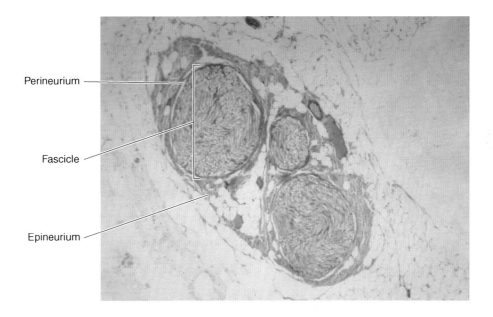

Perineurium

Fascicle

Epineurium

PHOTO 375b ◆ Peripheral nerve cross section (40x)

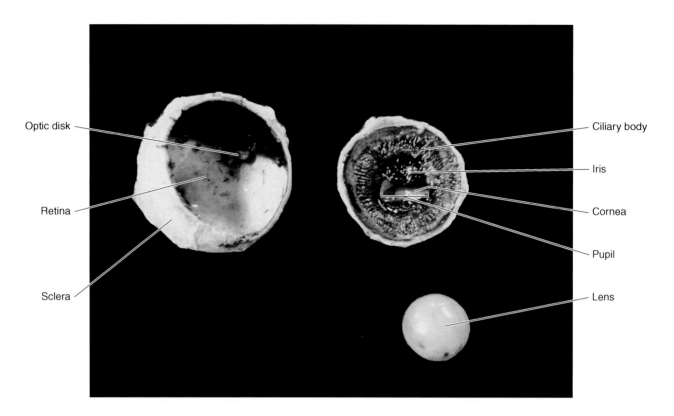

Optic disk

Retina

Sclera

Ciliary body

Iris

Cornea

Pupil

Lens

PHOTO 376a ◆ Dissected cow eye

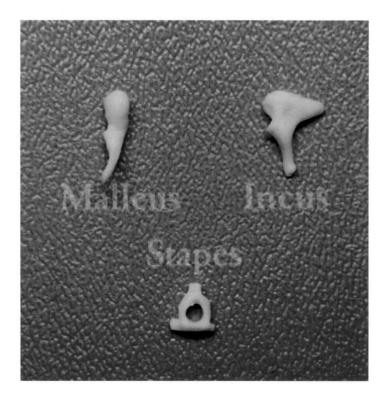

PHOTO 376b ◆ Ossicles

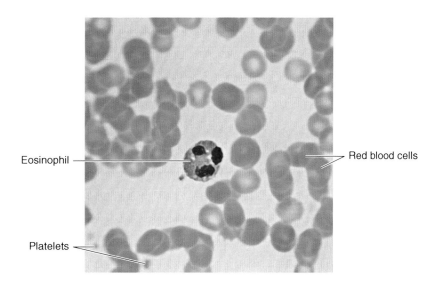

Eosinophil

Red blood cells

Platelets

PHOTO 377a ◆ Eosinophils (1000x)

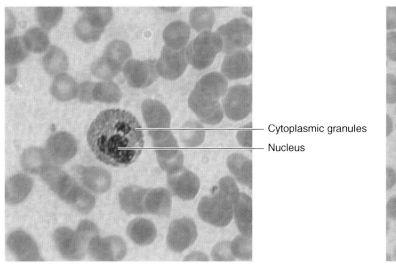

Cytoplasmic granules

Nucleus

PHOTO 377b ◆ (1000x)

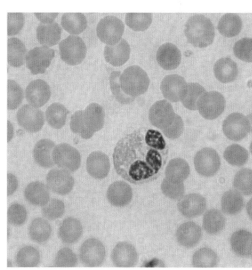

PHOTO 377c ◆ (1000x)

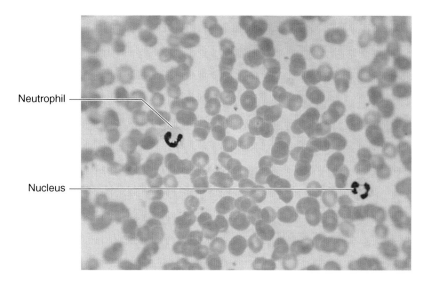

Neutrophil

Nucleus

PHOTO 378a ◆ Neutrophils (1000x)

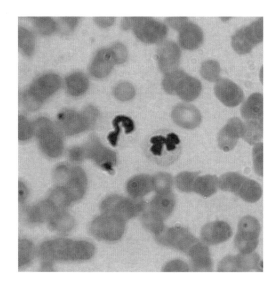

PHOTO 378b ◆

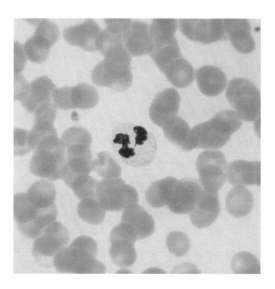

PHOTO 378c ◆

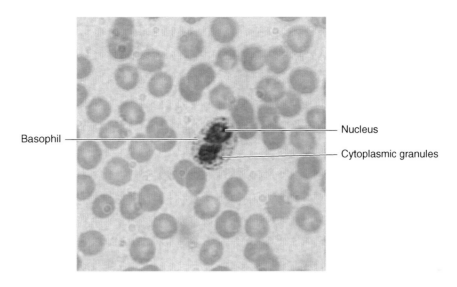

Basophil ————————

———— Nucleus

———— Cytoplasmic granules

PHOTO 379a ◆ Basophils (1000x)

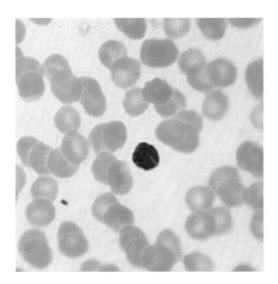

PHOTO 379b ◆

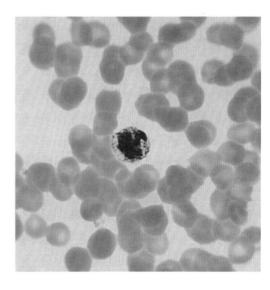

PHOTO 379c ◆

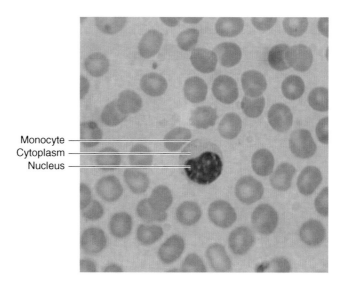

Monocyte ——————
Cytoplasm ——————
Nucleus ——————

PHOTO 380a ◆ Monocytes (1000x)

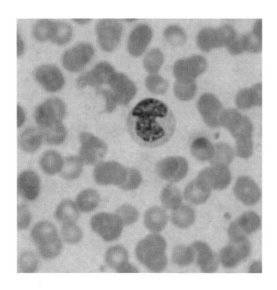

PHOTO 380b ◆

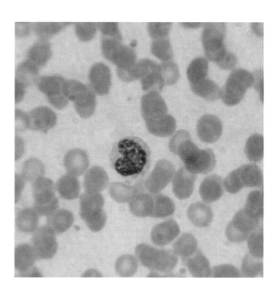

PHOTO 380c ◆

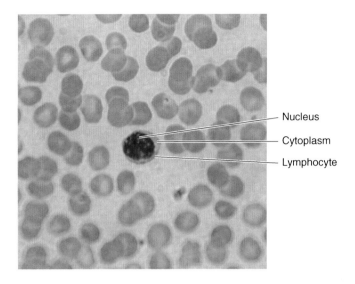

PHOTO 381a ◆ Lymphocytes (1000x)

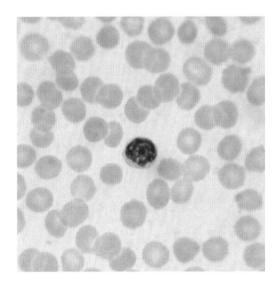

PHOTO 381b ◆

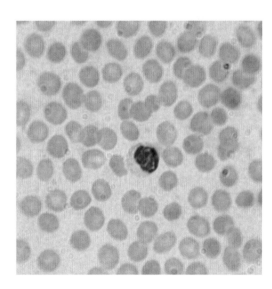

PHOTO 381c ◆

Aortic
arch

Brachiocephalic
trunk

Left common
carotid artery

Left subclavian
artery

Ascending aorta

Superior vena
cava

Left auricle

Left coronary
artery

Right auricle

Right coronary
artery

Marginal artery

Small cardiac
vein

Anterior
interventricular
artery and Great
cardiac vein

Left circumflex artery
in atrioventricular sulcus

Apex

Left pulmonary
veins

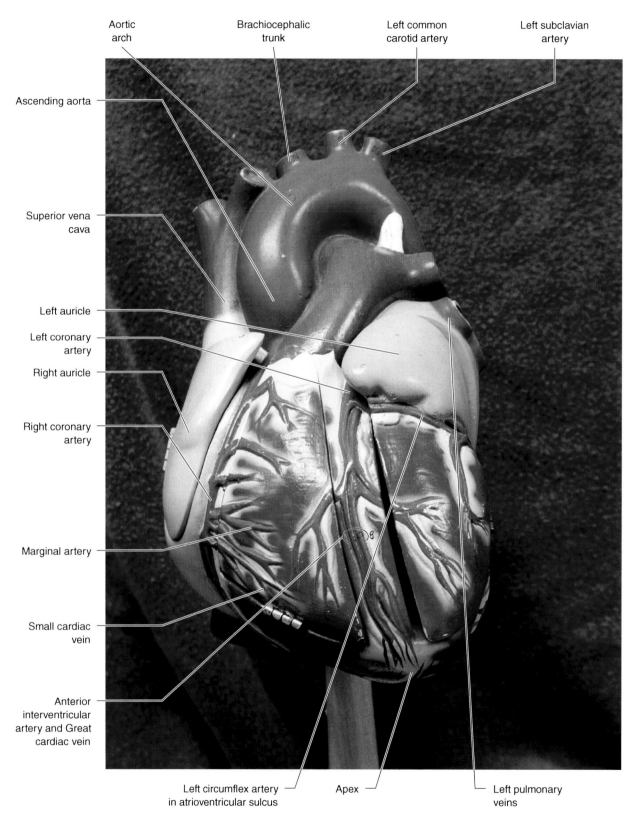

PHOTO 382 ◆ Heart model

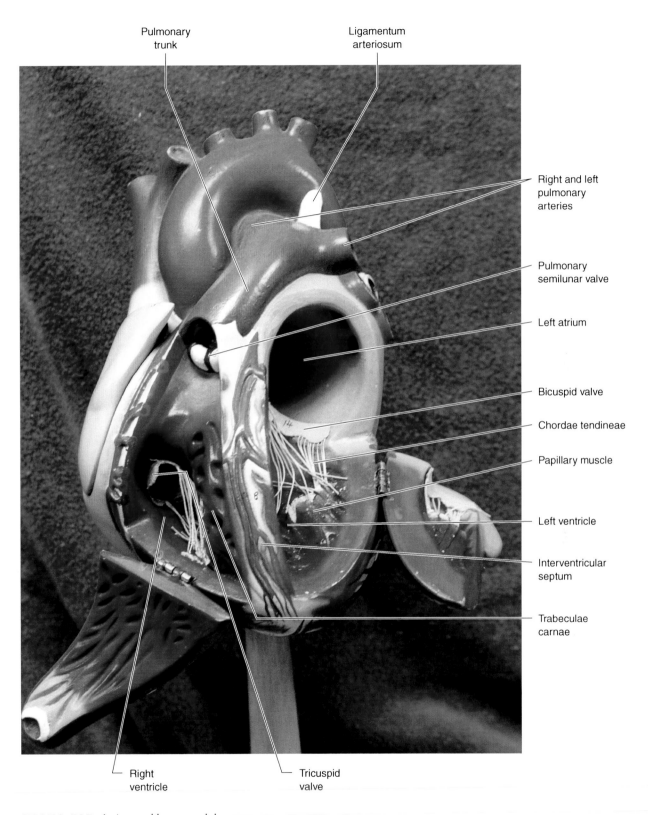

Pulmonary trunk

Ligamentum arteriosum

Right and left pulmonary arteries

Pulmonary semilunar valve

Left atrium

Bicuspid valve

Chordae tendineae

Papillary muscle

Left ventricle

Interventricular septum

Trabeculae carnae

Right ventricle

Tricuspid valve

PHOTO 383 ◆ Internal heart model

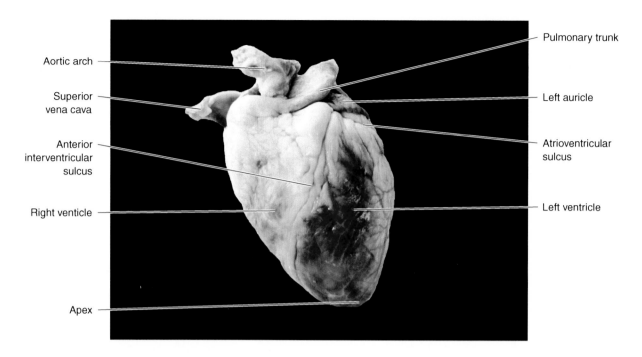

Aortic arch

Superior
vena cava

Anterior
interventricular
sulcus

Right venticle

Apex

Pulmonary trunk

Left auricle

Atrioventricular
sulcus

Left ventricle

PHOTO 384a ◆ Sheep heart external

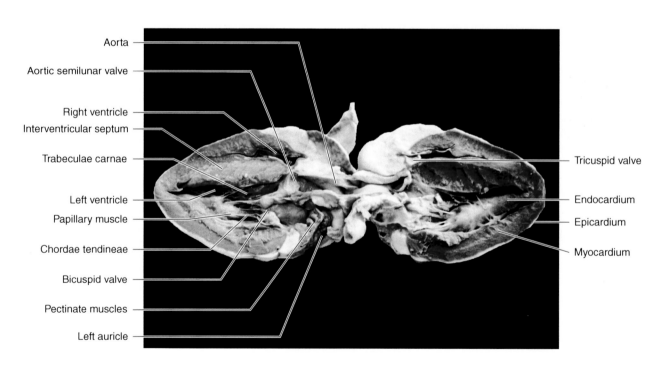

Aorta

Aortic semilunar valve

Right ventricle
Interventricular septum

Trabeculae carnae

Left ventricle
Papillary muscle

Chordae tendineae

Bicuspid valve

Pectinate muscles

Left auricle

Tricuspid valve

Endocardium

Epicardium

Myocardium

PHOTO 384b ◆ Sheep heart internal

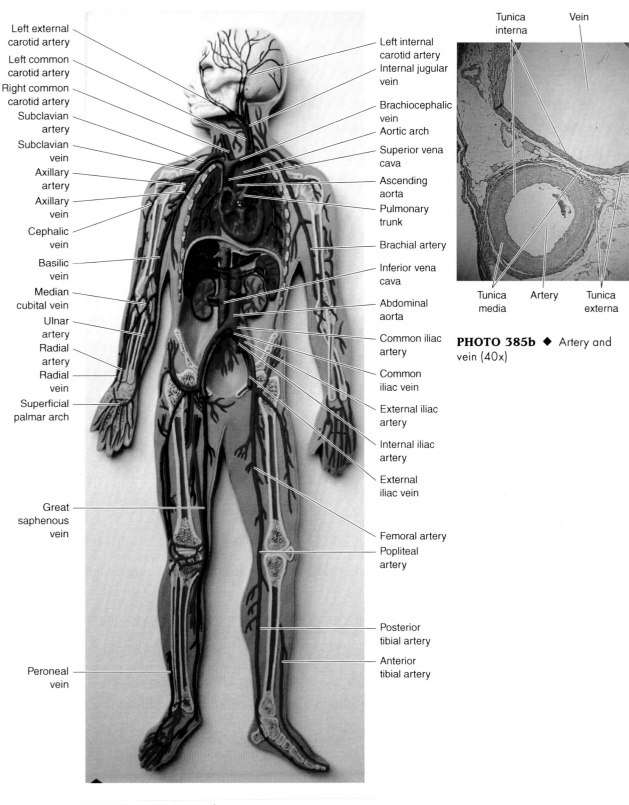

Left external
carotid artery

Left common
carotid artery

Right common
carotid artery

Subclavian
artery

Subclavian
vein

Axillary
artery

Axillary
vein

Cephalic
vein

Basilic
vein

Median
cubital vein

Ulnar
artery

Radial
artery

Radial
vein

Superficial
palmar arch

Great
saphenous
vein

Peroneal
vein

Left internal
carotid artery

Internal jugular
vein

Brachiocephalic
vein

Aortic arch

Superior vena
cava

Ascending
aorta

Pulmonary
trunk

Brachial artery

Inferior vena
cava

Abdominal
aorta

Common iliac
artery

Common
iliac vein

External iliac
artery

Internal iliac
artery

External
iliac vein

Femoral artery

Popliteal
artery

Posterior
tibial artery

Anterior
tibial artery

PHOTO 385a ◆ Circulatory system

Tunica
interna

Vein

Tunica
media

Artery

Tunica
externa

PHOTO 385b ◆ Artery and
vein (40x)

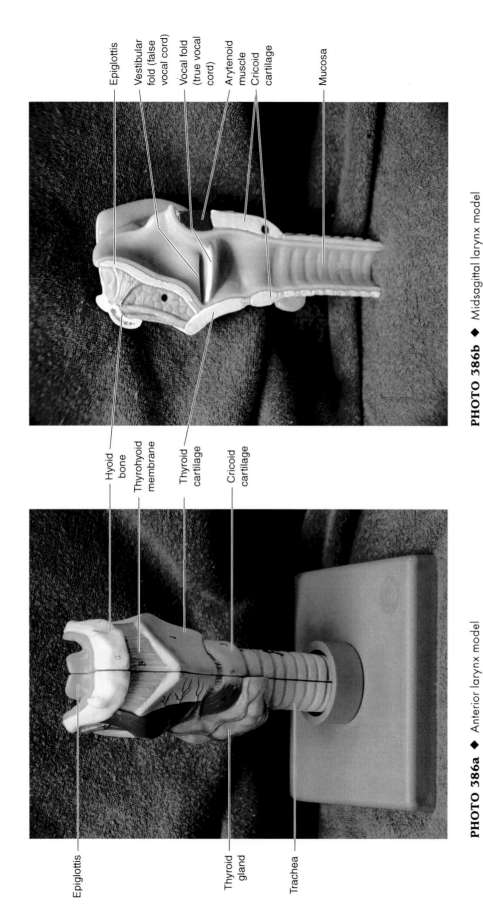

Epiglottis

Vestibular fold (false vocal cord)

Vocal fold (true vocal cord)

Arytenoid muscle

Cricoid cartilage

Mucosa

PHOTO 386b ◆ Midsagittal larynx model

Hyoid bone

Thyrohyoid membrane

Thyroid cartilage

Cricoid cartilage

Epiglottis

Thyroid gland

Trachea

PHOTO 386a ◆ Anterior larynx model

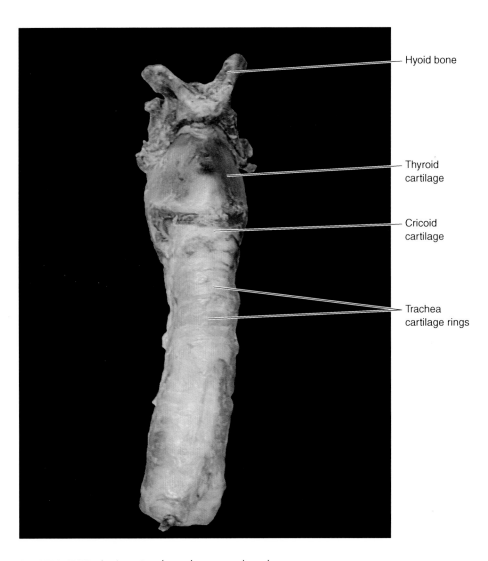

Hyoid bone

Thyroid
cartilage

Cricoid
cartilage

Trachea
cartilage rings

PHOTO 387 ◆ Anterior sheep larynx and trachea

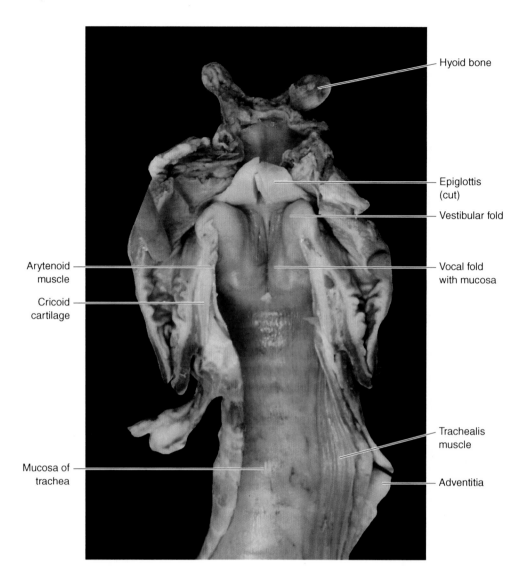

Hyoid bone

Epiglottis
(cut)

Vestibular fold

Vocal fold
with mucosa

Arytenoid
muscle

Cricoid
cartilage

Trachealis
muscle

Adventitia

Mucosa of
trachea

PHOTO 388 ◆ Posterior view of internal sheep larynx
and trachea

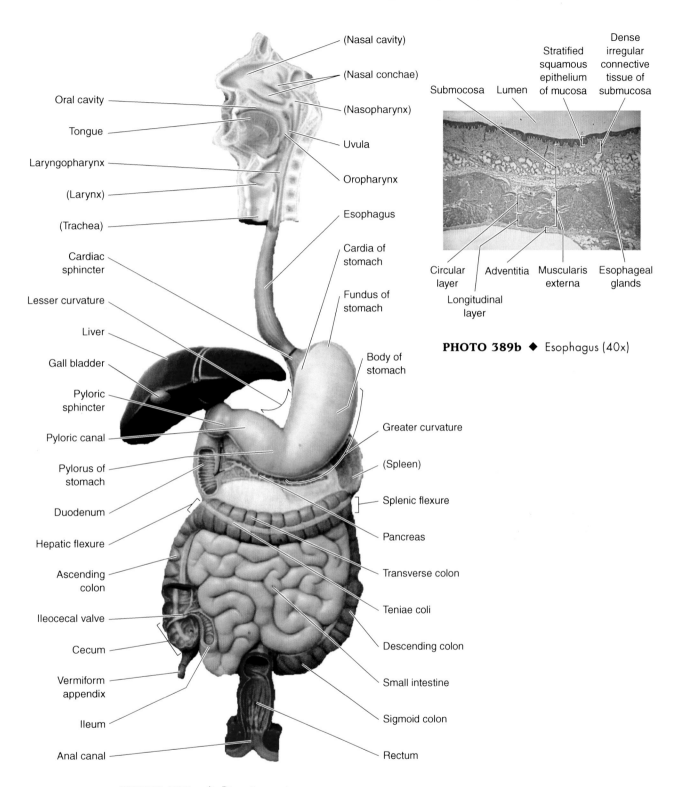

Oral cavity

Tongue

Laryngopharynx

(Larynx)

(Trachea)

Cardiac sphincter

Lesser curvature

Liver

Gall bladder

Pyloric sphincter

Pyloric canal

Pylorus of stomach

Duodenum

Hepatic flexure

Ascending colon

Ileocecal valve

Cecum

Vermiform appendix

Ileum

Anal canal

(Nasal cavity)

(Nasal conchae)

(Nasopharynx)

Uvula

Oropharynx

Esophagus

Cardia of stomach

Fundus of stomach

Body of stomach

Greater curvature

(Spleen)

Splenic flexure

Pancreas

Transverse colon

Teniae coli

Descending colon

Small intestine

Sigmoid colon

Rectum

PHOTO 389a ◆ Digestive system

Submocosa   Lumen

Stratified squamous epithelium of mucosa

Dense irregular connective tissue of submucosa

Circular layer

Longitudinal layer

Adventitia

Muscularis externa

Esophageal glands

PHOTO 389b ◆ Esophagus (40x)

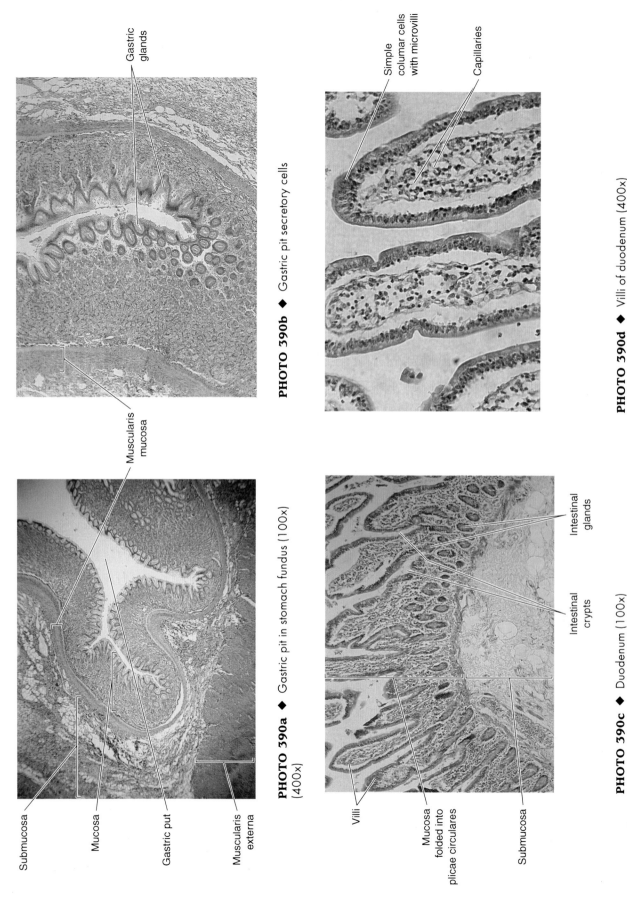

PHOTO 390a ◆ Gastric pit in stomach fundus (100x)

PHOTO 390b ◆ Gastric pit secretory cells (400x)

PHOTO 390c ◆ Duodenum (100x)

PHOTO 390d ◆ Villi of duodenum (400x)

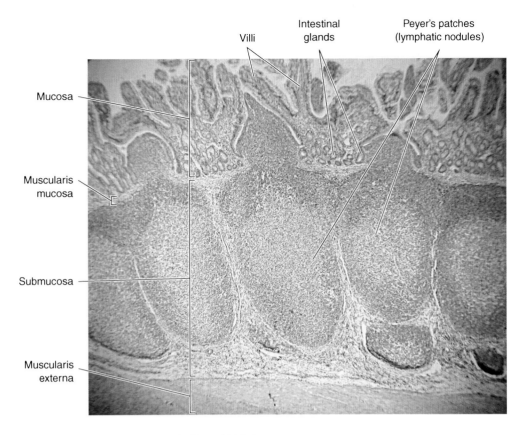

Villi

Intestinal
glands

Peyer's patches
(lymphatic nodules)

Mucosa

Muscularis
mucosa

Submucosa

Muscularis
externa

PHOTO 391a ◆ Ileum (100x)

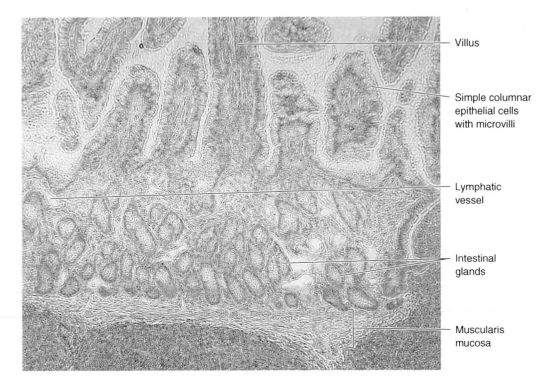

Villus

Simple columnar
epithelial cells
with microvilli

Lymphatic
vessel

Intestinal
glands

Muscularis
mucosa

PHOTO 391b ◆ Ileum mucosa (400x)

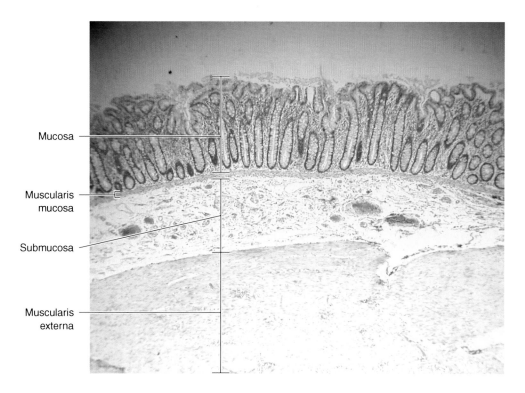

Mucosa

Muscularis
mucosa

Submucosa

Muscularis
externa

PHOTO 392a ◆ Colon (100x)

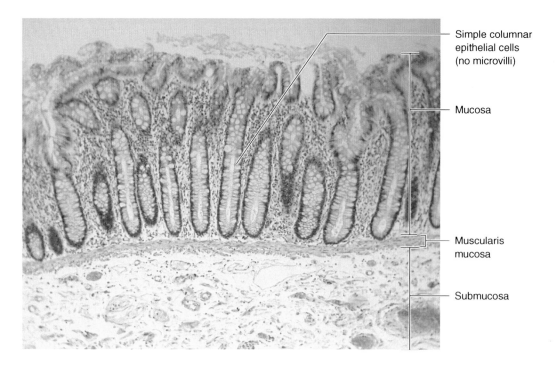

Simple columnar
epithelial cells
(no microvilli)

Mucosa

Muscularis
mucosa

Submucosa

PHOTO 392b ◆ Colon (400x)

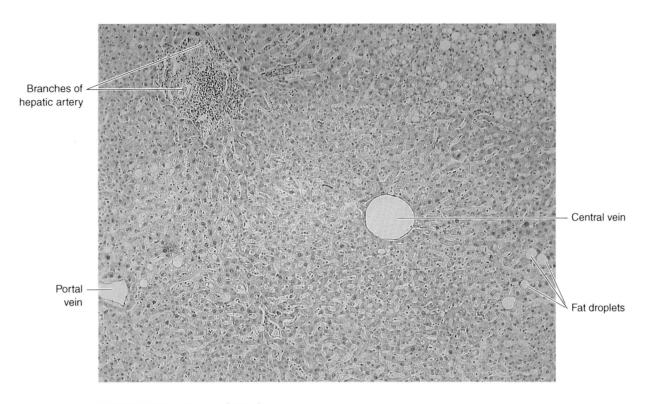

Branches of
hepatic artery

Central vein

Portal
vein

Fat droplets

PHOTO 393a ◆ Liver (100x)

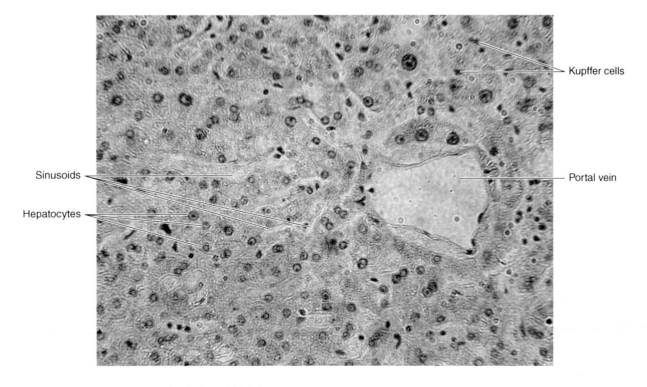

Kupffer cells

Sinusoids

Portal vein

Hepatocytes

PHOTO 393b ◆ Liver (400x)

Proximal
convoluted
tubule

Descending
limb of loop
of Henle

Arcuate artery
and vein

Renal
medullary
pyramid

Lobar
artery

Segmental
artery

Renal
artery

Renal
vein

Hilus

Renal
pelvis

Ureter

Distal
convoluted
tubule

Ascending
limb of loop
of Henle

Collecting
duct

Renal
cortex

Glomeruli

Interlobar
artery and
vein

Interlobular
artery and
vein

Major
calyx

Minor
calyces

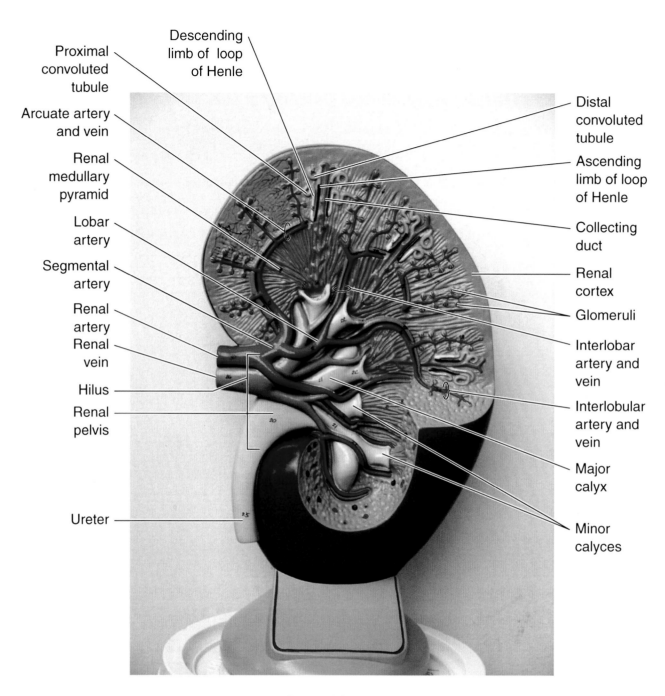

PHOTO 394 ◆ Kidney model

Renal capsule

Interlobular artery

Interlobular vein

Interlobar artery

Interlobar vein

Arcuate vein

Lobar artery

Renal column

Papilla

Medullary pyramid

Cortex

Major calyx

Minor calyx

Hilus

Renal pelvis

Ureter

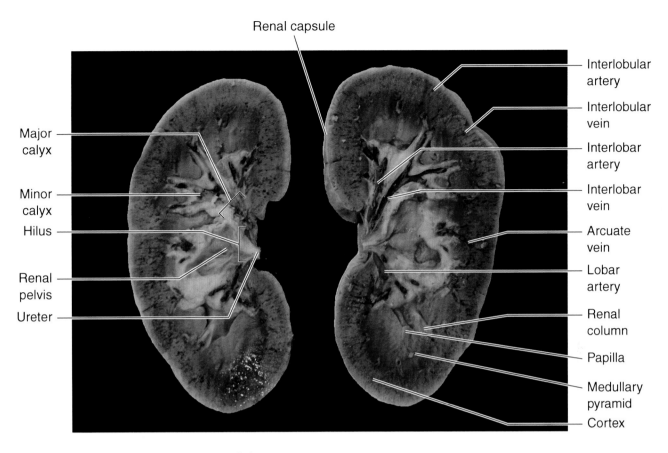

PHOTO 395 ◆ Pig kidney

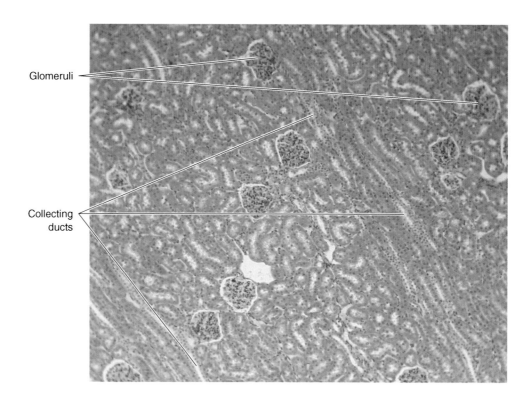

Glomeruli

Collecting
ducts

PHOTO 396a ◆ Renal cortex (100x)

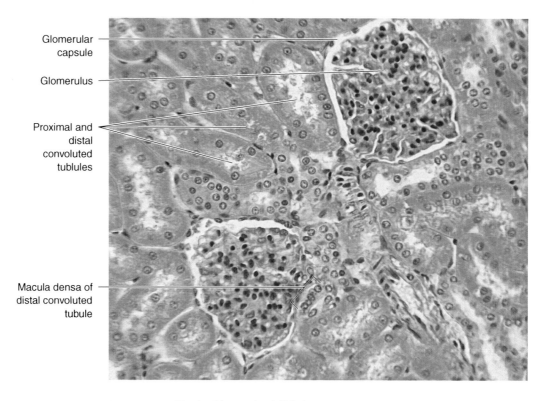

Glomerular
capsule

Glomerulus

Proximal and
distal
convoluted
tublules

Macula densa of
distal convoluted
tubule

PHOTO 396b ◆ Glomerulus (400x)

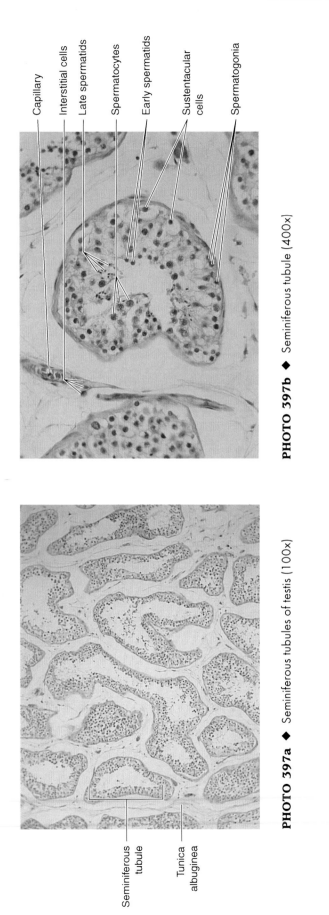

PHOTO 397b ◆ Seminiferous tubule (400x)

PHOTO 397a ◆ Seminiferous tubules of testis (100x)

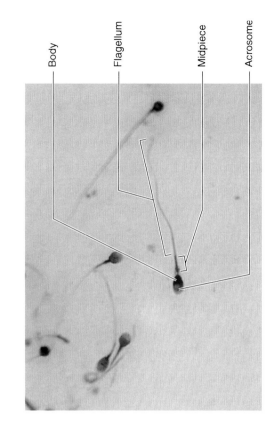

PHOTO 397d ◆ Human sperm cell

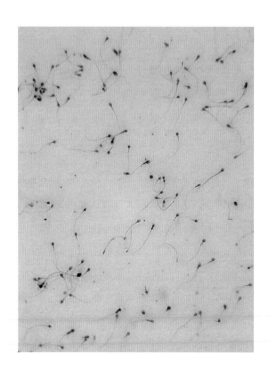

PHOTO 397c ◆ Human spermatazoa (400x)

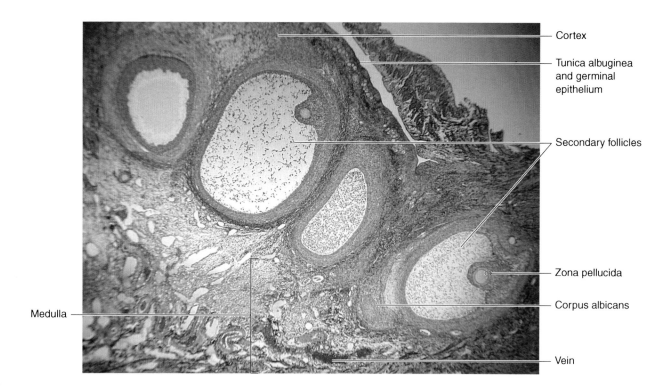

Cortex

Tunica albuginea
and germinal
epithelium

Secondary follicles

Zona pellucida

Corpus albicans

Medulla

Vein

PHOTO 398a ◆ Ovary (40x)

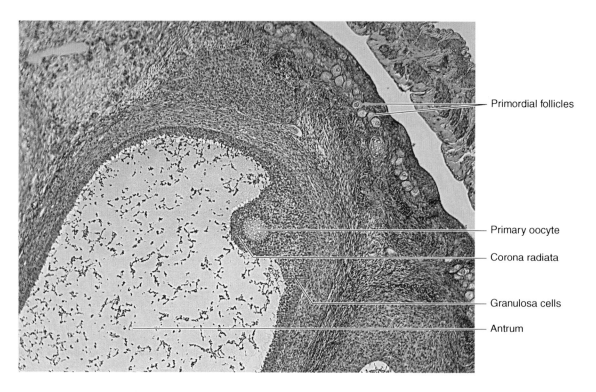

Primordial follicles

Primary oocyte

Corona radiata

Granulosa cells

Antrum

PHOTO 398b ◆ Follicle in ovary (100x)

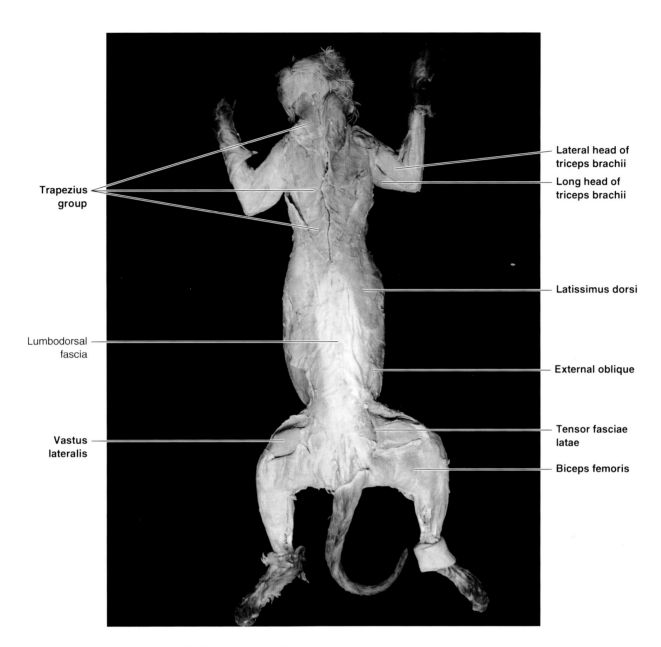

Trapezius group

Lateral head of triceps brachii

Long head of triceps brachii

Latissimus dorsi

Lumbodorsal fascia

External oblique

Vastus lateralis

Tensor fasciae latae

Biceps femoris

PHOTO 399 ◆ Dorsal cat muscles

Pectoantebrachialis

Pectoralis major

Pectoralis minor

Serratus ventralis (anterior)

Latissimus dorsi reflected

External oblique

linea alba

Rectus abdominis

Sartorius

Gracilis

Semitendinosus

Gastrocnemius

Tibialis anterior

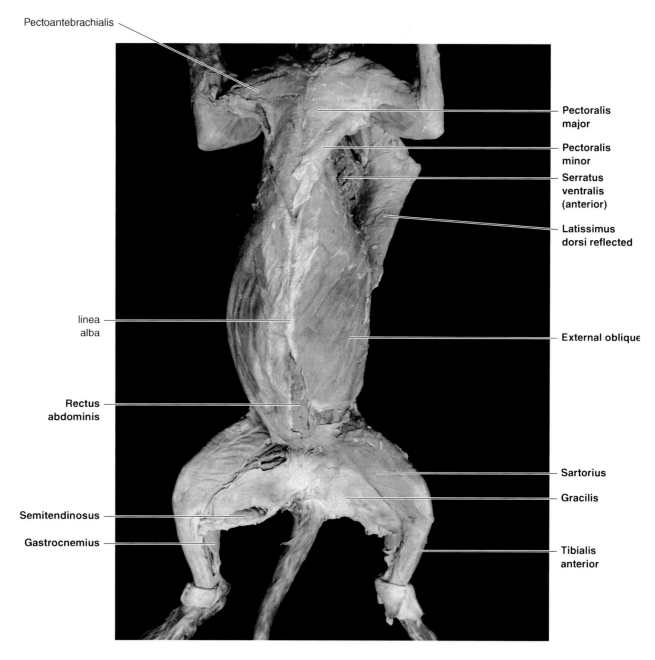

PHOTO 400 ◆ Ventral cat muscles

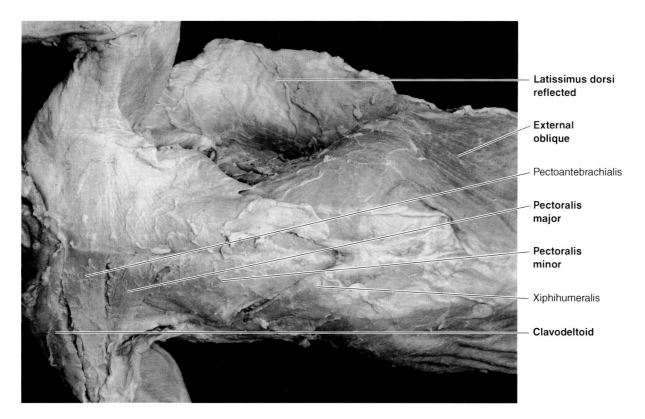

PHOTO 401a ◆ Muscles of ventral thorax

Latissimus dorsi reflected

External oblique

Pectoantebrachialis

Pectoralis major

Pectoralis minor

Xiphihumeralis

Clavodeltoid

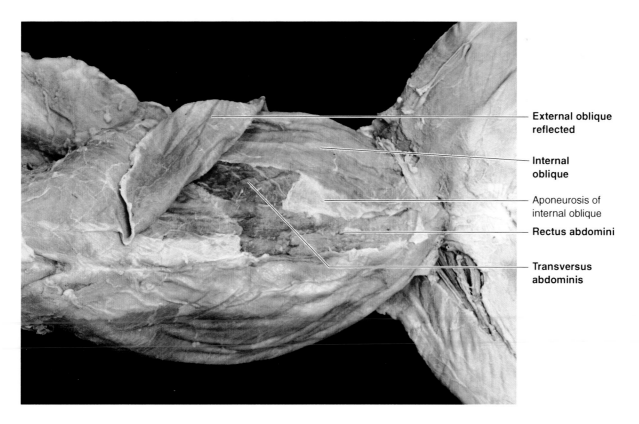

PHOTO 401b ◆ Abdominal muscles

External oblique reflected

Internal oblique

Aponeurosis of internal oblique

Rectus abdomini

Transversus abdominis

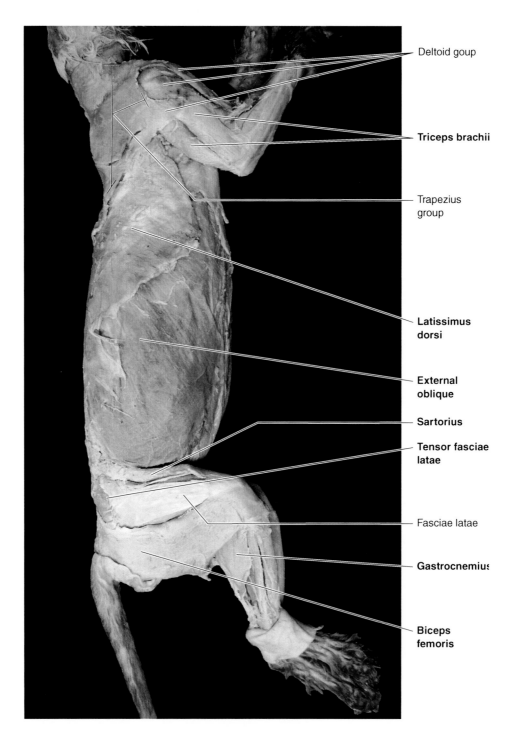

Deltoid goup

Triceps brachii

Trapezius
group

**Latissimus
dorsi**

**External
oblique**

Sartorius

**Tensor fasciae
latae**

Fasciae latae

Gastrocnemius

**Biceps
femoris**

PHOTO 402 ◆ Lateral view of cat muscles

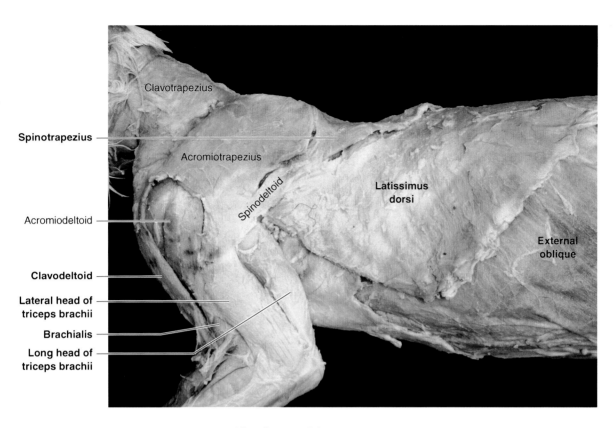

Clavotrapezius

Spinotrapezius

Acromiotrapezius

Spinodeltoid

Latissimus
dorsi

Acromiodeltoid

External
oblique

Clavodeltoid

**Lateral head of
triceps brachii**

Brachialis

**Long head of
triceps brachii**

PHOTO 403a ◆ Lateral brachium and thorax

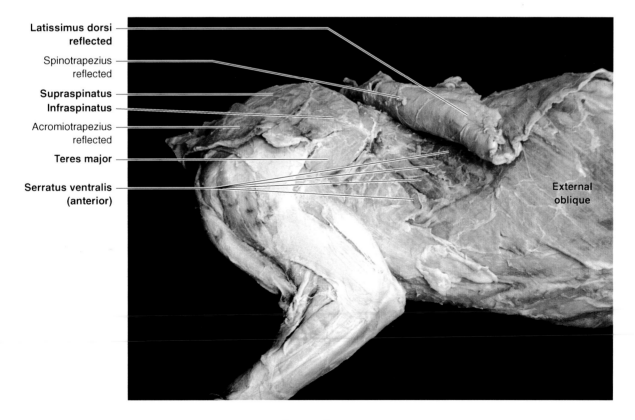

**Latissimus dorsi
reflected**

Spinotrapezius
reflected

Supraspinatus

Infraspinatus

Acromiotrapezius
reflected

Teres major

**Serratus ventralis
(anterior)**

External
oblique

PHOTO 403b ◆ Deep scapular muscles

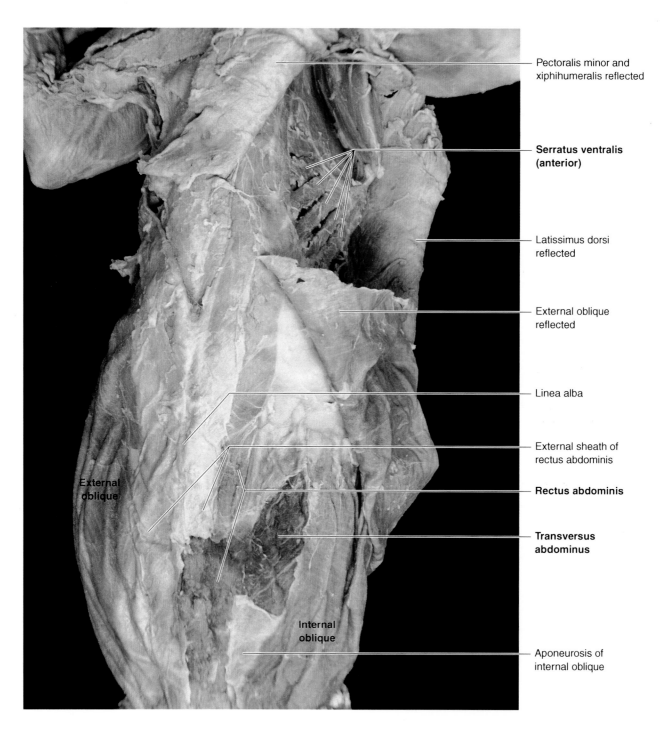

Pectoralis minor and
xiphihumeralis reflected

**Serratus ventralis
(anterior)**

Latissimus dorsi
reflected

External oblique
reflected

Linea alba

External sheath of
rectus abdominis

Rectus abdominis

**Transversus
abdominus**

Aponeurosis of
internal oblique

External
oblique

Internal
oblique

PHOTO 404 ◆ Deep ventral muscles of thorax and abdomen

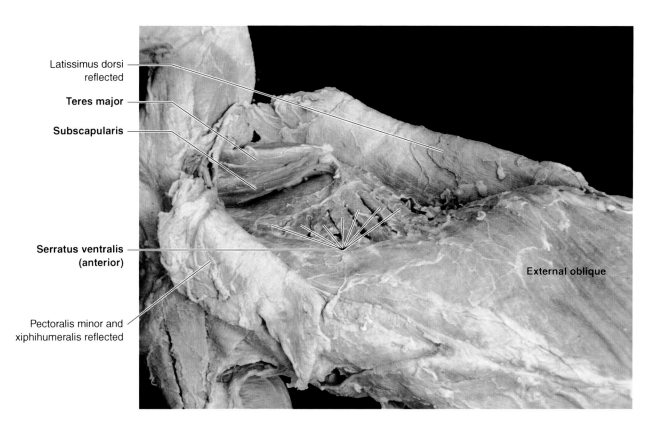

Latissimus dorsi
reflected

Teres major

Subscapularis

**Serratus ventralis
(anterior)**

Pectoralis minor and
xiphihumeralis reflected

External oblique

PHOTO 405a ◆ Deep muscles of ventral scapula

Rhomboideus

Rhomboideus capitus

Acromiotrapezius
reflected

Supraspinatus

Infraspinatus

Sinotrapezius and
latissimus dorsi
reflected

PHOTO 405b ◆ Deep muscles of dorsal scapula

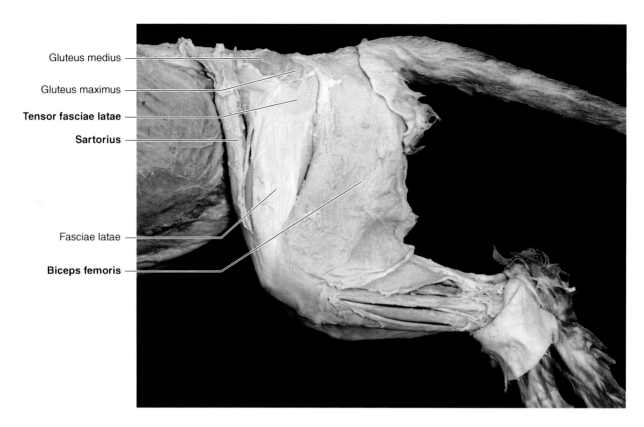

Gluteus medius

Gluteus maximus

Tensor fasciae latae

Sartorius

Fasciae latae

Biceps femoris

PHOTO 406a ◆ Superficial muscles of lateral leg

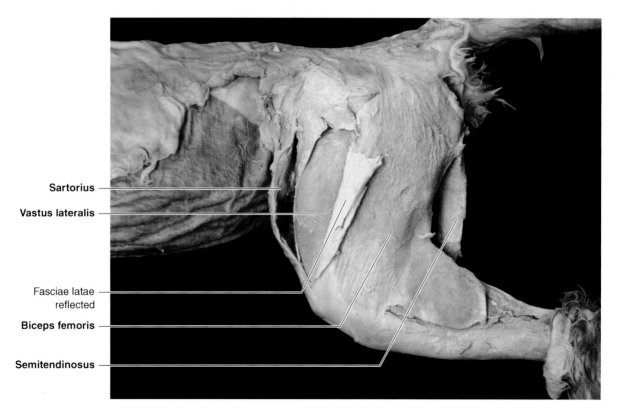

Sartorius

Vastus lateralis

Fasciae latae reflected

Biceps femoris

Semitendinosus

PHOTO 406b ◆ Reflected fascia lata

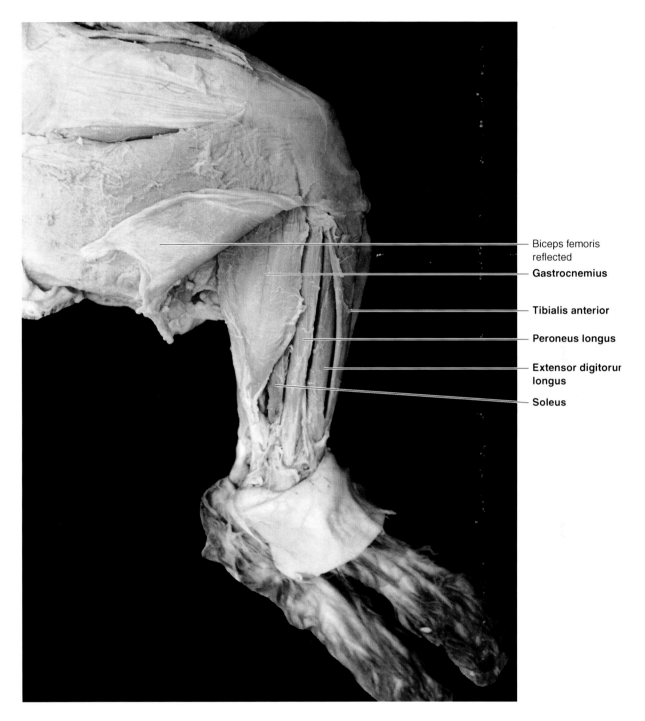

Biceps femoris
reflected

Gastrocnemius

Tibialis anterior

Peroneus longus

**Extensor digitorur
longus**

Soleus

PHOTO 407 ◆ Muscles of sural and peroneal regions

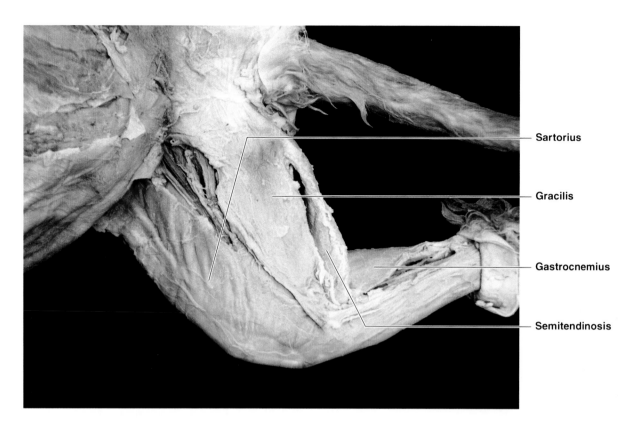

Sartorius

Gracilis

Gastrocnemius

Semitendinosis

PHOTO 408a ◆ Superficial muscles of medial leg

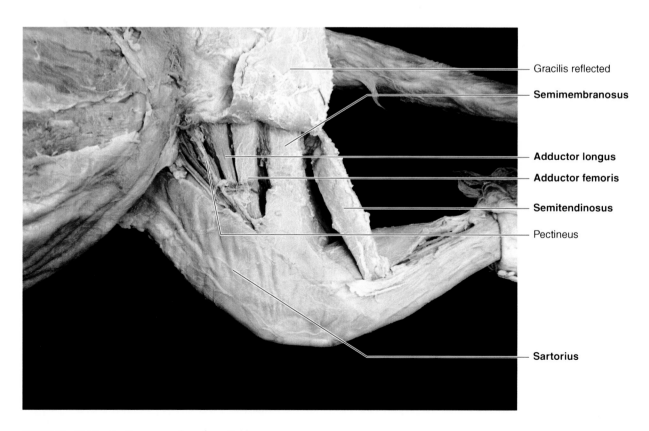

Gracilis reflected

Semimembranosus

Adductor longus

Adductor femoris

Semitendinosus

Pectineus

Sartorius

PHOTO 408b ◆ Deep muscles of medial leg

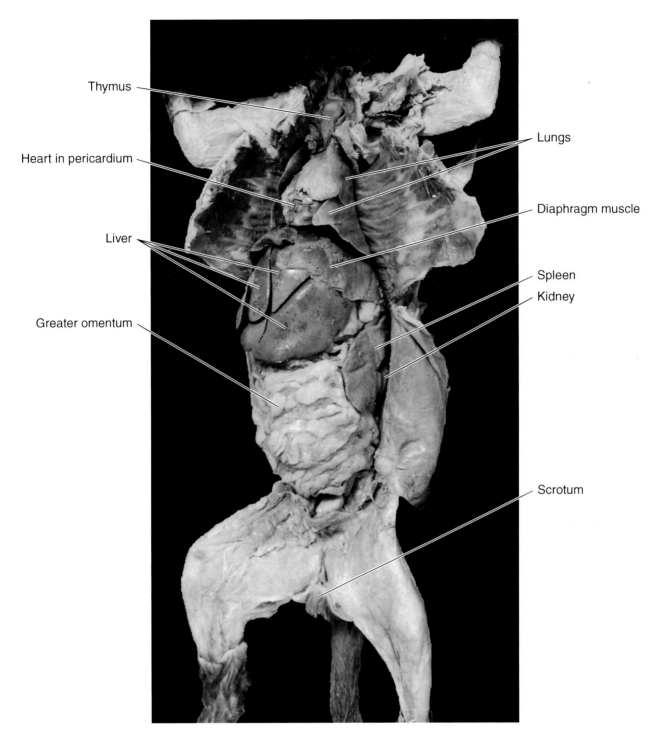

Thymus

Heart in pericardium

Liver

Greater omentum

Lungs

Diaphragm muscle

Spleen

Kidney

Scrotum

PHOTO 409 ◆ Thoracic and abdominal viscera with greater omentum in place

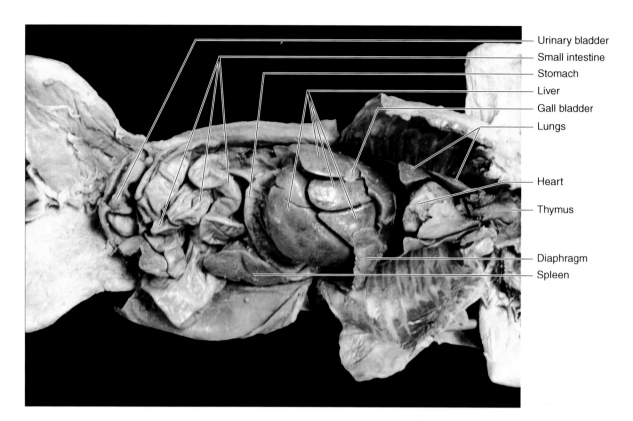

Urinary bladder
Small intestine
Stomach
Liver
Gall bladder
Lungs

Heart

Thymus

Diaphragm
Spleen

PHOTO 410a ◆ Thoracic and abdominal viscera (greater omentum removed)

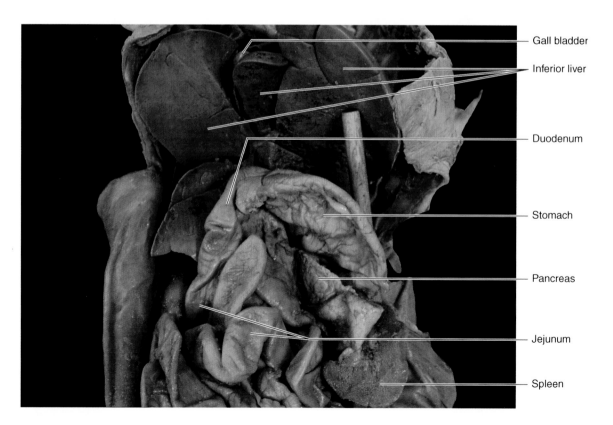

Gall bladder

Inferior liver

Duodenum

Stomach

Pancreas

Jejunum

Spleen

PHOTO 410b ◆ Inferior liver and digestive viscera (greater omentum removed)

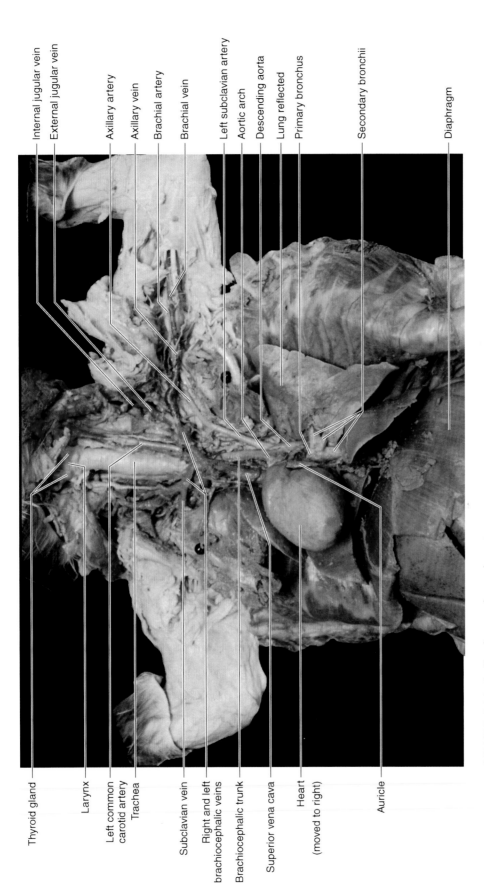

Internal jugular vein
External jugular vein
Axillary artery
Axillary vein
Brachial artery
Brachial vein
Left subclavian artery
Aortic arch
Descending aorta
Lung reflected
Primary bronchus
Secondary bronchii
Diaphragm

Thyroid gland
Larynx
Left common carotid artery
Trachea
Subclavian vein
Right and left brachiocephalic veins
Brachiocephalic trunk
Superior vena cava
Heart (moved to right)
Auricle

PHOTO 411 ◆ Cardiovascular and respiratory organs

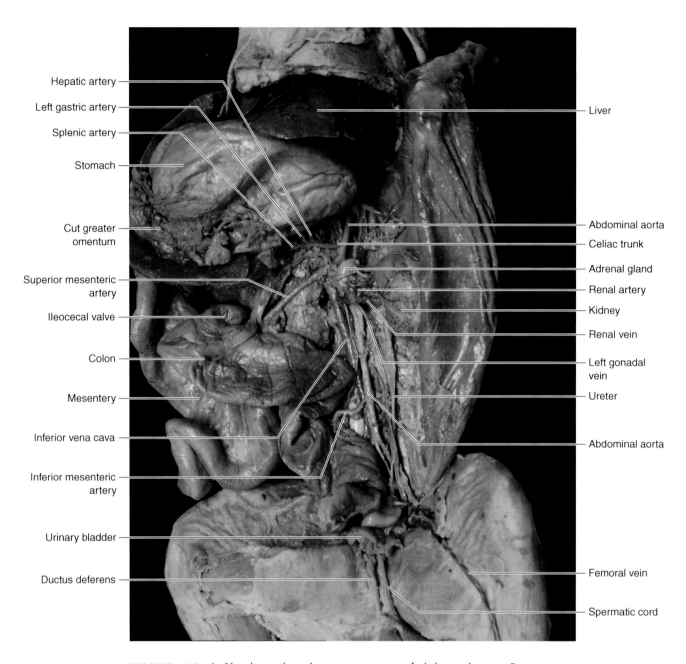

Hepatic artery

Left gastric artery

Splenic artery

Stomach

Cut greater
omentum

Superior mesenteric
artery

Ileocecal valve

Colon

Mesentery

Inferior vena cava

Inferior mesenteric
artery

Urinary bladder

Ductus deferens

Liver

Abdominal aorta

Celiac trunk

Adrenal gland

Renal artery

Kidney

Renal vein

Left gonadal
vein

Ureter

Abdominal aorta

Femoral vein

Spermatic cord

PHOTO 412 ◆ Blood vessels and urinary structures of abdominal cavity. Digestive viscera and spleen moved to right side.

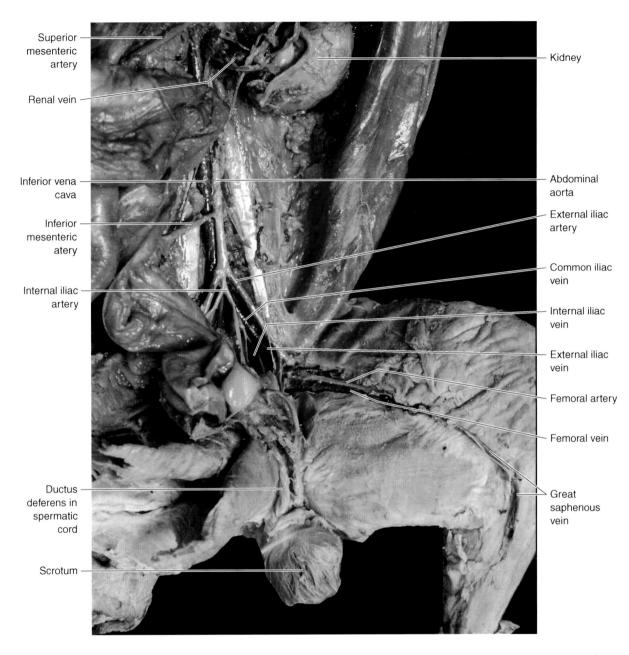

Superior mesenteric artery

Renal vein

Inferior vena cava

Inferior mesenteric atery

Internal iliac artery

Ductus deferens in spermatic cord

Scrotum

Kidney

Abdominal aorta

External iliac artery

Common iliac vein

Internal iliac vein

External iliac vein

Femoral artery

Femoral vein

Great saphenous vein

PHOTO 413 ◆ Blood vessels of inguinal regions. Intestines moved to right side.

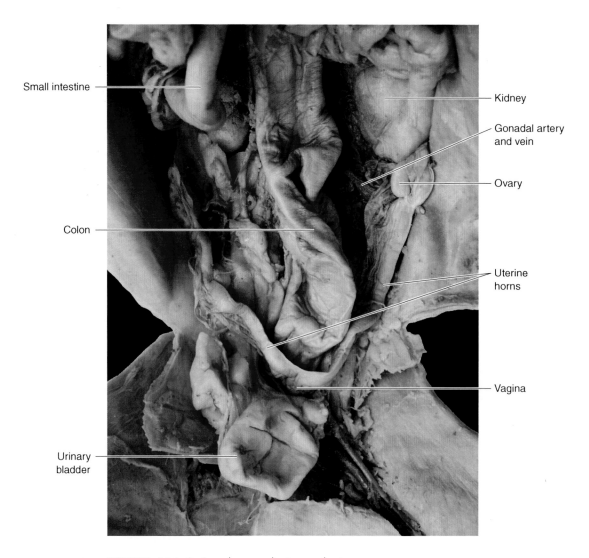

Small intestine

Kidney

Gonadal artery
and vein

Ovary

Colon

Uterine
horns

Vagina

Urinary
bladder

PHOTO 414 ◆ Female reproductive and urinary organs